The Successful Internship

TRANSFORMATION AND EMPOWERMENT

H. Frederick Sweitzer
University of Hartford

Mary A. King
Fitchburg State College

Brooks/Cole Publishing Company

I(T)P® An International Thomson Publishing Company

Pacific Grove • Albany • Belmont • Bonn • Boston • Cincinnati • Detroit
Johannesburg • London • Madrid • Melbourne • Mexico City • New York
Paris • Singapore • Tokyo • Toronto • Washington

Sponsoring Editor: *Eileen Murphy*
Marketing Team: *Steve Catalano, Margaret Parks*
Editorial Assistant: *Julie Martinez*
Production Editor: *Keith Faivre*
Manuscript Editor: *Frank Hubert*
Permissions Editor: *Connie Dowcett*
Design Editor: *Roy R. Neuhaus*
Interior Design & Interior Illustration: *TBH/Typecast, Inc.*
Cover Design: *Susan Horovitz*
Typesetting: *TBH/Typecast, Inc.*
Printing and Binding: *Webcom*

CREDITS: **Page 7:** List of questions from "Reaching Out: School-Based Community Service Programs" by National Crime Prevention Council, 1988, p. 101. Reprinted with permission. **Page 9:** Table 1.1 from *Human Services: Contemporary Issues and Trends* by H. S. Harris & D. C. Maloney. Copyright © 1996 by Allyn and Bacon. Reprinted by permission. **Page 56:** Exercise from *A Handbook for Developing Multicultural Awareness* by Paul Pederson. Copyright © 1988 American Counseling Association. Adapted with permission. **Page 191:** "Ethical Standards of Human Services Professionals" In *Human Service Education, 16*(1), by the National Organization for Human Service Education, pp. 11–17. Reprinted with permission.

For more information, contact:

BROOKS/COLE PUBLISHING COMPANY
511 Forest Lodge Road
Pacific Grove, CA 93950
USA

International Thomson Publishing Europe
Berkshire House 168–173
High Holborn
London WC1V 7AA
England

Thomas Nelson Australia
102 Dodds Street
South Melbourne, 3205
Victoria, Australia

Nelson Canada
1120 Birchmount Road
Scarborough, Ontario
Canada M1K 5G4

International Thomson Editores
Seneca 53
Col. Polanco
11560 México, D. F., México

International Thomson Publishing GmbH
Königswinterer Strasse 418
53227 Bonn
Germany

International Thomson Publishing Asia
60 Albert St.
#15-01 Albert Complex
Singapore 189969

International Thomson Publishing Japan
Hirakawacho Kyowa Building, 3F
2-2-1 Hirakawacho
Chiyoda-ku, Tokyo 102
Japan

Printed in Canada.

10 9 8 7 6 5 4 3 2

Library of Congress Cataloging-in-Publication Data

Sweitzer, H. Frederick.
The successful internship : transformation and empowerment / H. Frederick Sweitzer, Mary A. King.
p. cm.
Includes bibliographical references and index.
ISBN 0-534-35782-2
1. Interns—United States—Handbooks, manuals, etc. 2 Internship programs—United States. 3. Experiential learning—United States. 4. Human services—Study and teaching (Higher)—United States.
I. King, Mary A. II. Title.
LC1072.I58S84 1999
371.2'27 — dc21 98-27609
CIP

THIS BOOK IS DEDICATED, WITH LOVE, TO OUR MOTHERS

From Fred,
to my first writing teacher, my staunchest supporter,
and my biggest fan, Elizabeth C. Sweitzer.

From Mary,
to Phyllis Sadowski King,
who taught me the importance of lessons from the heart.

ABOUT THE AUTHORS

H. Frederick Sweitzer is Associate Professor of Human Services at the University of Hartford in Connecticut, where he also serves as Chair of the Division of Education in the College of Education, Nursing and Health Professions. Fred has over 20 years' experience in human services as a social worker, administrator, teacher, and consultant. He has placed and supervised undergraduate interns for 14 years and developed the internship seminar at the University of Hartford. Fred brings to his work a strong background in self-understanding, human development, experiential education, and group dynamics. He is on the editorial board for the journal *Human Service Education* and has published widely in the field.

Mary A. King is Professor of Human Services at Fitchburg State College in Massachusetts, where she is Chair of the Graduate Program in Criminal Justice and coordinates the Certificate Program in Forensic Casework. She teaches in the human services, criminal justice, and counseling programs and supervises graduate and undergraduate interns. Mary coordinated field placements for these and other programs for over 15 years and has published several articles on internship issues. She brings to her work a background in public education, juvenile justice, and private practice. Mary is licensed in mental health counseling, marriage and family therapy, and social work.

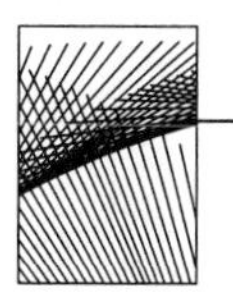

Contents

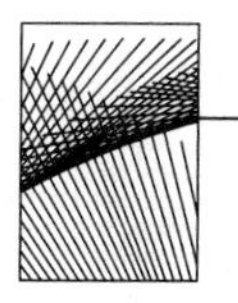

Foreword

While this book is written for human services interns, with some adaptation it will be extremely useful to interns in any field. It is well structured to aid those faculty new to the field, while encouraging experienced internship advisors to reexamine the rigor of their supervision of interns. There are the materials one would expect in such a text—helping students understand themselves, their values, motivations; walking through experiential education theory; and creating a learning plan or contract. But the sections on supervisory style, relationships with co-workers, and understanding the organization of the placement I think cover ground not usually dealt with and will be particularly helpful. If students use the text and its accompanying exercises responsibly, *they* will have the tools to make their internships powerful learning and that is the heart of experiential education.

Anne Kaplan
Faculty Emeritus
Former Editor, *NSEE Quarterly*

Preface

An internship can be a very powerful experience. Near the end of a course of study, students devote 15, 20, 30, or even more hours each week to working in an actual work setting in their chosen or prospective field. It is a time when skills are sharpened, when various aspects of the students' preparation are synthesized and integrated into one experience, and when theory and practice come together in a way that enhances both.

As instructors associated with interns for many years, and before that as agency personnel working with interns from many disciplines, we have felt the power of that experience. We look forward each year to our work with interns; we learn something every time and we never tire of the process.

Nevertheless, there always seem to be some struggles involved with an internship. To some extent, struggle is a necessary and important part of any internship, but we each felt for some time that the struggles could be fewer in number, milder in intensity, and more easily resolved if they were handled differently. Most important of all, we wanted the struggles to be opportunities for learning. Over time, we tried to anticipate the sorts of issues that would arise and to facilitate the reflection that is such a crucial part of experiential learning. And that is the purpose of this book.

The Successful Internship is intended for two main audiences. Since the early 1970s, human services has gained an identity as a helping profession separate from, but related to, other helping professions such as counseling and social work. Its goal is to prepare generalists, who have a broad base of skills and can function in a wide variety of settings. It has its own philosophy, its own professional organizations, its own professional journals, and its own code of ethics. Bachelor's and associate's programs in the United States, Canada, and Australia prepare students in the human services philosophy. Practically every such program has an internship component. Human services is also a generic term used as an umbrella to include many specific programs in areas such as mental health, criminal justice, and even social work. Many of these programs use internships as well.

We hope this book will be of use to all these audiences and that students, faculty, and field supervisors will find it helpful. It is intended to be used as part of a seminar class that accompanies an internship experience. It is possible to use it as an independent guide or workbook, but it was written for instructors and students to use together. It could also be used as a text for a site-based seminar conducted for interns at a particular setting.

We have intended the book to be more phenomenological than skill-based. It is not about the terminology, skills, and knowledge that interns need to work with a particular client population or in a particular system. Because of the wide variety of settings in which interns function, any such attempt would miss the mark for a significant number of students. We also assume that a good deal of the professional skills and knowledge needed for the internship have been covered elsewhere in the academic program. Rather, we have focused on a separate set of skills and knowledge designed to help interns make sense of the *experience* of the internship and of the emotions and reactions they go through every day, every week. We have tried to accomplish this in a way that is compatible with many different theoretical orientations to human service practice, while not specifically targeting any one of them.

The book is organized around two "big ideas." The first is that there are some issues and concerns that interns encounter at certain stages in their internships. Interns progress through these stages in a reasonably predictable order, though not at any predictable rate of speed. The second big idea is the importance of self-understanding. In addition to knowing about clients, about techniques and methods, and about organizations and administration, students need to know something about themselves. We believe this makes them better interns and better practitioners. It also helps them recognize the stages because individual interns experience the stages in their own unique ways.

With these big ideas in mind, we open the book with a section that gives students a thorough grounding in self-understanding, especially concerning issues relevant to the internship, and a brief introduction to the five stages of an internship. As the book progresses through the sections and chapters, each of the stages is discussed. We also consider the issues and tasks that must be dealt with if students are to move successfully to the next stage. The theme of self-understanding is reiterated throughout so that students can see connections between what they have learned about themselves and the shape and pace of their experience of the internship and its stages.

We have drawn on many theories in writing this book, including some of our own, but we have tried to keep the actual theoretical material to a minimum. When we believe a theory is important, we have tried to give the reader enough information to understand its value and application to the internship. We have also provided resources for further exploration. Faculty may want to use these resources for their own background knowledge, and individual instructors may also choose to emphasize some theories more than others. We have tried to write a book that allows for that flexibility.

There are questions and activities for structured reflection at the end of each chapter. Some of these exercises will have greater meaning than others for a particular intern or group of interns, or for particular instructors. Therefore, we do not recommend that all of them be routinely assigned but rather that instructors and students work together to select those questions and activities most relevant and thought-provoking for them.

Our experience has been that students go through the stages in the order we discuss, but often at different rates of speed. In our seminar classes, we do cover the chapters in order and at a particular time in the internship, and that seems to work for the majority of the group. However, we also listen carefully to what individual students are saying in class and in their journals, and we often suggest that a particular student reread a chapter or skip ahead to future ones as needed.

ACKNOWLEDGMENTS

This book has been a labor of love. Our relationship to the ideas that make up the book, to the task of writing it, and to one another has evolved as any relationship does. And, like any relationship, the book has evolved in the context of support, encouragement and guidance from many sources. Both of us have enjoyed sabbatical leaves from our institutions that helped us jump-start the process of writing. We appreciate having been given that time. Our colleagues from a number of disciplines have listened to our ideas, expanded our thinking with their ideas, and buoyed us with their enthusiasm. Some have even used portions of the manuscript in their work with interns. We are grateful for their willingness to contribute to our book in those ways and we thank in particular Robert Fried, John Hancock, Mary Ann Hanley, Anita Hotchkiss, Margot Kempers, and Norma Kronenberg.

The National Organization for Human Service Education, the New England Organization of Human Services Education, and the National Society for Experiential Education have been our professional homes and have provided us with forums to present and discuss our ideas. We thank our many colleagues in those organizations who have attended workshops, shared ideas, and reviewed and published our work. Our special thanks go to Gary Hesser, Will Holton, Anne Kaplan, Pat Kenyon, Tricia McClam, Ralph C. Meyer, Lynn McKinney, Anita Runyan, and Marianne Woodside, who have shown particular interest in and support for our work.

We both studied at the University of Massachusetts at Amherst, and at the Summer Clinical Developmental Institutes at Harvard University. We had wonderful teachers who left a lasting impression on us. Many of them are cited directly in the book, and all of them have had a substantial impact on the way we see the world and our work. We thank in particular Gerald Weinstein, Robert Kegan, Jack Wideman, Bailey Jackson, Rita Hardiman, Don Carew, Rene Carew, and Alan Ivey.

Many reviewers contributed their time and expertise to making this a better book. Our heartfelt thanks go to Robert Fried, Univerity of Hartford; Tricia McClam, University of Tennessee; Lynn McKinney, Univerity of Rhode Island; Susan Membrino, Assumption College; Punky Pletan-Cross, LUK Crisis Center; and Vicki Gardine Williams, Ferris State University for their thoughtful reviews, astute insights, and valuable suggestions. Our students have worked with various iterations of this manuscript and have not been shy about telling us what works and what does not. Their voices and all they have taught us over the years are evident throughout the pages of this book.

We are grateful as well for the support of agency supervisors who have cared about and for our students, for their insight and their unswerving commitment to quality internships. We especially thank Jesse Balcom, Tim Blair, Oseh Cole, Laura Gailbraith, Lucian Manzi, Richard Peterson, Jerry Reisman, and Robert White. We are also thankful for the technical and research assistance provided by Linda Cone and Janice Ouelette of the library staff at Fitchburg State College and by Eileen Ganong, Cynthia Hill, Tricia May, and Joan MacDonald. Their work was far from glamorous, but their contributions were invaluable.

Our warm appreciation goes to the staff at Brooks/Cole. There would have been no book without Claire Verduin, who believed we had something important to say and waited patiently while we found the time and energy to launch the project. Eileen Murphy took over the book after Claire's retirement and maintained her enthusiasm and patience throughout the project. The production team did a superb job of turning the manuscript into a book and were a pleasure to work with. We are especially grateful to production editor Keith Faivre, design editor Roy Neuhaus, copyeditor Frank Hubert, and interior designer Bill Turner.

Our families and friends have supported us, put up with us, and made do without us through the long process of writing. Family members Charlie King, Skip, Betty, and Sally Sweitzer, Britt Howe, Phyllis Agurkis, Joe and Michael King, and Heidi Sadowski, deserve special mention. So, too, do our dear friends Jeff and Judy Bauman, Vicky Day, Steve Eisenstat and Vivian Page, Regina Miller, Julie Portman and Paul Reisler, Sam Pukitis, Kathy Rondeau, and Ed Weinswig. Finally, special thanks and love go to our life partners, Martha Sandefer and Peter Zimmermann, and to Mary's son Patrick, who was only a dream when this project was launched. We are grateful to be finished before his adolescence.

SECTION ONE

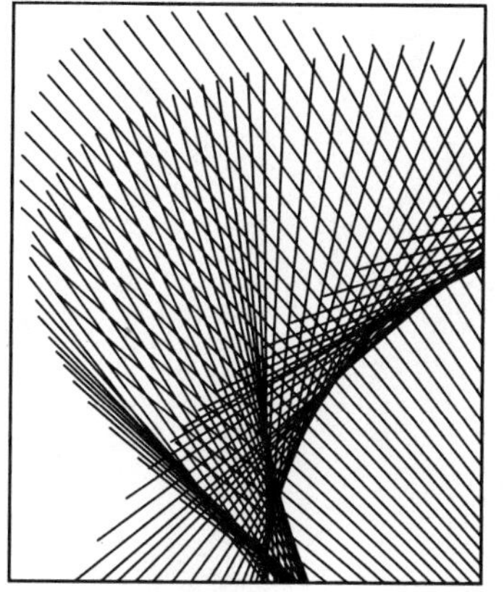

LAYING THE GROUNDWORK

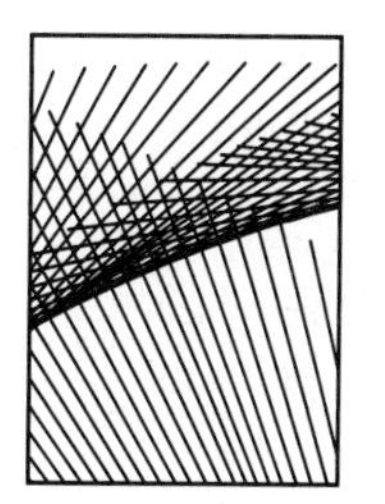

CHAPTER 1

Introduction and Overview

Education is revelation that affects the individual.
GOTTHOLD EPHRAIM LESSING, 1780

It gives meaning to everything you have learned and makes practical sense of something you've known as theoretical.
STUDENT JOURNAL ENTRY

WELCOME TO YOUR INTERNSHIP

You are beginning what is, for most students, the most exciting experience of your education program. Chances are you have looked forward to this experience for a long time. You've probably heard your share of stories—both good and bad—from other, more experienced students. And while you may be a small minority on your campus, you join virtually thousands of other students all over the country. An intensive field experience is a critical component of many education programs—including psychology, sociology, and criminal justice—and of almost every human service education program (E. Simon, 1989). Many human services students conduct their internships in a social service setting; others choose corporate or research settings.

In addition to internships required by particular majors, there are increasing numbers of colleges and universities requiring or encouraging an intensive community service experience regardless of major. The number of hours required varies widely, from as few as 80 hours to as many as 500 (E. Simon, 1989). There are many terms for these experiences, including internship, field work, field experience, co-op education, field education, and practicum. We will use the term *internship* to refer to those experiences that involve receiving academic credit for working in a social service agency, or at some other site, for at least 8 hours per week over the course of a semester.

The internship is the culminating experience for many students; it occurs near the end of the academic program and is a chance to pull together and apply much of what you have learned. Many students see the internship as a chance to (finally) learn to actually *do* something. Although you may have had courses that developed particular skills, the internship is a chance to improve those skills and acquire lots of new ones. Skill development, though, is only one of the possible goals and outcomes of your experience. It is a chance to apply theory to practice. Actually, it is a chance to develop the relationship between theory and practice, for each should inform the other. The theories you have studied (and continue to study) should help you analyze and perform effectively in various situations. However, your experience will also help you see where the theories do not quite apply or where you need to search for a new theoretical model to help you.

The internship also affords you the opportunity to understand the world of work in a more complete way than you do now. Even if you have had full-time jobs, presumably your internship is taking you into an area in which you have little professional experience. The internship is also a catalyst for personal growth. If you give yourself a chance, you can learn a tremendous amount about yourself. Finally, the internship can help you clarify your career and educational goals.

So Why Do You Need a Book?

The internship is a learning experience like no other. In a classroom learning experience, you learn through readings, lectures, discussions, and exercises. These are the raw materials you are given. You bring your ability to memorize, analyze, synthesize, and evaluate to these materials—these are your learning tools. The arena for learning is the classroom, and classrooms vary in the amount and quality of interaction between students and the instructor, and among students. Some can be very interactive places, with mutual dialogue among students as well as between students and teachers. Other classrooms are interactive, but only between teacher and student; it is almost as if there is a multitude of individual relationships being carried on in isolation. The internship is different in every way, and this book is designed to help you and your instructor with this new approach.

Furthermore, the internship is not just an intellectual experience. It is a *human* experience, full of all the wonderful and terrible feelings that people bring to their interactions and struggles. And your clients are not the only ones having these feelings; you will have them too. This emotional, human side of the internship is not just a backdrop to the real work and the real learning; it is every bit as real and important.

Experience, both intellectual and emotional, is the raw material of the internship. You will be learning mostly through experience, although you may engage in some traditional academic activities as well. However, one noted theorist in the field of experiential education, David Kolb (1984), suggests that experience alone does not lead to learning or growth. Rather, the experience must be processed and organized in some way. You must think about your experience, sometimes in structured ways, and discuss it with others. Reflective dialogue with yourself and your peers is the primary tool for learning in the internship. This book is designed to help you structure your reflection and dialogue. It will invite you to think about your internship in a variety of ways, some of which may be new to you. It may also help you anticipate some of the challenges that await you and move successfully through them.

Relationships are the medium of the internship; they are the context in which most of your learning and growth occurs. Many of you will have clients at your placement, but even if you do not, you will be involved in relationships with a supervisor at your placement, an instructor on campus, other interns on site and on campus, and co-workers at your placement. These relationships offer rich and varied opportunities for learning and growth. This book will help you think about these relationships, capitalize on the opportunities they present, and address any problems that may arise.

In addition to excitement and satisfaction, most interns also experience some real difficulties—moments when they question themselves, their career choices, their placements, or all of the above. We like to think of these moments as crises, but not in the way you are probably familiar with the term. We prefer to think of crises as the Chinese do. The symbol for crisis in Chinese is a combination of two symbols: danger and opportunity. So, while there is some risk and certainly some discomfort in these challenges, there is also tremendous opportunity for growth. We hope this book will help you see both the dangers and the opportunities inherent in an internship and grow from both. We will encourage you not to run from these crises, but to meet them head on, with your mind and your heart open to the experience.

The truth is that some of you may not need a book; you will learn all you can and survive the crises just fine. However, we believe that for most students, this book, in combination with a skilled instructor, supportive peers, and some of your time and energy, can add to your learning.

Some Basic Terms

Although internships and community service experiences exist at many colleges and universities, different language is often used to describe the various aspects of the experience and the people associated with it. For example, the term *supervisor* sometimes refers to a person employed by the placement site and sometimes to a faculty or staff member at the college. So to avoid confusion, here are some terms we will be using and what they mean.

Placement or Site This term refers to the place where the student is interning, and sites can vary quite a bit. It could be a social service agency, a corporate setting, a college or university office, a hospital, or a school. Through the process of finding an internship, you probably are aware of the incredible variety of opportunities that exist in the community. If not, and if you are curious, there are books mentioned at the end of this chapter that you can consult.

Intern This is the term that refers to you, the student who is working at the site, even though you may not be called an intern at your college or university.

Supervisor Your supervisor is the person assigned by the placement site to meet regularly with you, answer your questions, and give you feedback on your progress. Most placements assign one site supervisor to one intern, although in some cases, there may be more than one person fulfilling these functions. Some academic programs use the term *field instructor* to describe this person in order to emphasize the educational (as opposed to managerial) nature of the role.

Instructor This is the faculty or staff member at your school who oversees your placement. In some cases, it is the same person who helped you find the placement, but in other cases, it is someone else. This person may meet with you individually during the semester, visit you at the site, hold conferences with you and your supervisor, conduct a seminar class for interns, grade your performance, or all of the above.

Co-Worker This term refers to the other people who work at the placement, regardless of their title, status, or how much you interact with them. If there are other interns at the site, from your school or some other school, they are in the role of co-worker when you are at the placement site.

THE INTERNSHIP SEMINAR

Most of you will be meeting with an instructor and other interns on campus during the semester. We refer to these meetings as seminars. The word *seminar* comes from the Italian *seminare*, which means to sow or seed. The class sessions are a medium that is most helpful in the integration of intellectual and affective learning, encouraging new understanding and creative responses, and strengthening the effectiveness of interpersonal relationships (Williams, 1975).

A seminar may be a bit different from other classes you have taken. For example, one basic assumption of a seminar is that each person has something to contribute (Royse, Dhooper, & Rompf, 1996), unlike many classroom experiences where the assumption often is that only the teacher has something to contribute. A seminar class is one in which an exchange of ideas takes place, where information is shared, and mutual problems are discussed. It is also a forum for problem-centered learning. If everyone has something to contribute, everyone shares the responsibility for the success of the experience. You have additional responsibilities, then, but also additional benefits.

An effective seminar affords opportunities for reflective dialogue, support, the development of important relationships, and a variety of new learning experiences. Other benefits of a seminar include enhancement of individual learning, integration of cognition and affect, strengthening of humanistic values, and increased interpersonal effectiveness (King, 1995). The seminar is also an opportunity to learn how other agencies and placement sites operate and how they approach common challenges and problems. You may engage in many different activities as part of the seminar, including lectures, discussions, reflective assignments, journal writing, student presentations, and support groups (which may meet in or outside of class).

Keeping a Journal

> *I believe that classwork and journals are critical to internships because they allow support from peers, feedback from teachers, and reflection on your own work and feelings.*
>
> STUDENT JOURNAL ENTRY

Instructors may require interns to keep a journal of some kind. Even if it is not required in your setting, we strongly recommend that you keep one and record the events, thoughts, and feelings from each day that appear significant to you. Although it may occasionally seem like a chore, if you put time into it, journal keeping will give you a way to see yourself growing and changing. It also forces you to take time on a regular basis to reflect on what you are doing. Many of the quotes you have seen and will continue to see throughout this book are drawn from student journals.

If you have a disability that makes it impossible or difficult for you to write, or if writing does not come easily to you, your journal could be tape recorded instead. Your instructor can listen to the tape each week and respond to you on tape or in writing, whichever the two of you prefer. Of course, you will need to negotiate these arrangements with your instructor, but a little time and thought should yield a method that allows you to reflect comfortably on your experience and maintain a dialogue with your instructor.

We suggest that you write an entry after every day that you go to your internship. Although there is no "right" length for these entries, they should record what you did and saw that day, new ideas and concepts you were exposed to and how you can use them, and your personal thoughts and feelings about what is happening to you. It may be helpful to divide what you learn at an internship into four categories: (a) knowledge, (b) skills, (c) personal growth, and (d) career development.

Knowledge refers to things you know about; for example, you might learn the principles of behavior modification. Skills are things you know how to do; for example, you know how to set up a behavior management program. Personal growth refers to what you have learned about yourself and your attitudes, values, reaction patterns, and personality traits. Career development refers to what you are learning about the field of human services and your place in it. Try to include all these categories in your journal.

Many interns tell us they are afraid that there are going to be days when there is just nothing to say. Well, our experience is that you won't have that happen very often, but there may be some days when writing is difficult. For those days, here are some questions to consider, generated by a community service program (National Crime Prevention Council, 1988):

- What was the best thing that happened today at your site? How did it make you feel?
- What thing(s) did you like least today about your site?
- What compliments did you receive today and how did they make you feel?
- What criticisms, if any, did you receive and how did you react to them?
- How have you changed or grown since you began your work at this site? What have you learned about yourself and the people you work with?
- How does working at this site make you feel? Happy? Proud? Bored? Why do you feel this way?
- Has this experience made you think about possible careers in this field?
- What kind of new skills have you learned since beginning to work at this site? How might they help you?
- What are some of the advantages or disadvantages of working at this occupation?
- If you were in charge of the site, what changes would you make?

- How has your work changed since you first started? Have you been given more responsibility? Has your daily routine changed at all?
- What do you think is your main contribution to the site?
- How do the people you work with treat you? How does it make you feel?
- What have you done this week that makes you proud?
- Has this experience been a rewarding one for you? Why or why not?

Remember, your journal has to work for you, so you may also want to include other kinds of entries. You may want to write some notes or reactions to the chapters you read in this book. Throughout the book, we will be discussing issues and suggesting many activities and exercises. Your reactions to the issues we raise and your work with the exercises could also go into your journal. Some interns include examples of their work, materials from the site, pictures, newspaper articles about the placement site or about their client population, and any other materials that they think are relevant to the experience. A well-kept journal of this sort is a gold mine to be drawn on for years to come. It becomes a portfolio of the experience as well as a record of the journey.

Gerald Weinstein (1981; Weinstein, Hardin, & Weinstein, 1975) developed a method of reflecting on events that may be helpful with your journal. Take a moment at the end of the week to recall any events that stand out in your mind. Select one or two (they can be positive or negative). Divide a piece of paper into three columns. In the left-hand column, record each action taken by you or others during the event. Record only those things that you saw or heard, such as "she frowned," "he said thank you," or "they stomped out of the room." List them one at a time. Now, review the list and try to recall what you were thinking when the different actions occurred. When you recall something, enter it in the middle column, directly across from the event. For example, you may have been thinking "what did I do now?" when the people left the room. Finally, read the list again and try to recall what you were feeling at the time each action and thought occurred. Record what you recall in the right-hand column. For example, you may have felt embarrassed, confused, or angry when they walked out. Table 1.1 presents an example of this sort of analysis.

If you do decide to keep a journal, make sure you are very clear with your instructor, supervisor, and clients about the intent of the journal and issues of confidentiality. If your journal is for your personal use only, then there is no issue. You have full responsibility for its contents and for ensuring that what you write is for your eyes only. However, if you want, or are required, to show it to other interns, your instructor, your supervisor, or anyone else, you must be careful not to disclose information about clients, the placement, or even yourself that is supposed to be kept private. Discuss this issue with your instructor and your site supervisor before going too far with your journal. You may also be concerned that you cannot be completely candid in your journal if some of the people you are writing about are going to read it! Some interns keep their journals in loose-leaf format and merely remove any pages they wish to keep private before showing the journal to anyone else.

Perhaps the most important thing you can do for your journal is to allot sufficient time to do it. Doing it over lunch on the due date is not a good approach! As you plan your days and weeks, leave at least 30 minutes after each day at your internship to write your daily entry and allow more time if you are including exercises from this book.

TABLE 1.1
Keeping a Journal: Reflecting on Events

Actions	Thoughts	Feelings
I am sitting in the lounge with several residents. John walks in and sits down. There are several chairs available, but he sits right in front of me.	This guy is always looking for trouble. What is he doing?	Nervous. Uncomfortable.
I say "hello." He nods. I continue my conversation with the residents.		
John squirms around in his chair several times. Finally, I notice the outline of a pack of cigarettes in his pants pocket (a clear violation of house rules).	What is his problem? Damn. He has cigarettes and I'm supposed to take them away and take points. He has a terrible temper. He set this whole thing up.	Annoyed. Angry. Anxious.
When I look up, he is looking right at me.	I have to do something now. He knows that I saw them.	Embarrassed. More anxious.
I say, "What've you got in your pocket there, John?"		
John: "Where?"	Here we go.	
Me: "Right there."		
John: "Nothing! What are you talking about?"	He's not going to make this easy. I'm trying to be nice.	Nervous and angry.
Me: "The cigarettes. You obviously wanted me to see them."		
John: "I did not! So what are you going to do about it anyway?"	If I punish him now, he's going to do something worse.	Confused. Uncertain.
Me: "What do you think I should do?"	I'm stalling and he knows it.	Stupid.
John: "I think you should leave me alone." His face is getting red.	I'm tired of this nonsense.	Angry. Resentful.
Me: "If you wanted that, you shouldn't have come in here. You could have just gone outside and smoked, you know."	I can't believe I just suggested he break a rule–I just wouldn't have had to deal with him if I hadn't seen him.	Upset at myself.
John: "Go take a flying leap [expletives deleted]!" He jumps to his feet.	Uh oh! Are there any other staff around? I have to calm him down. The other kids are watching me.	Scared. Self-conscious. Alert.
Me: "Look, if you just give me the cigarettes, I won't report this."	Maybe this will work.	Hopeful.

(continued)

TABLE 1.1 *(continued)*

Actions	Thoughts	Feelings
John: "They're mine. No one takes my property!" He is clenching his fists.	He's not going to get physical over this. Is he?	Frightened.
Me: "You're not allowed to have them here and you know it. I should take points away."		
John: "Do I get them back?"	I can't give them back.	Confused. Desperate.
Me: "I don't know. I'll think about it."		
John: "All right, but only because I like you."		
Me: "Thanks."	Thank God. I wonder if I did the right thing, though.	Relieved. Embarassed. Angry.

Support

It is important that interns have a place where they can talk about their experiences, their feelings and reactions, and their struggles and achievements. Although your friends and family can do some of this for you, it is often helpful to have this exchange with others who are undergoing a similar experience. Support groups exist for almost every purpose; perhaps you have participated in some. While the seminar is not a support group in the formal sense, one of its principal benefits is the quality of connections that you develop with your peers. Through these relationships, you give and receive support. In fact, you receive a double benefit; not only do you give and receive support, but you also become more skilled at each of these functions.

Effective support is both very simple and quite difficult, especially when someone is sharing a problem. Often, the person who is listening offers advice or speculates on the underlying psychological dynamics of the persons involved. If you have taken a helping skills class, you probably know that giving advice is often not the best approach, but when a friend or classmate is struggling, it can be awfully tempting to offer your heartfelt suggestions. We have found that it is usually best to hold your advice until and unless it is asked for. It may be helpful for someone to hear whether others have had a similar problem and how they handled it, but let the request come from that person.

In addition, remember that this is not a therapy group. You are not therapists (nor are your instructors in a therapist's role, although they may have those skills). There may be times when certain individuals encounter a challenge at the internship or an experience that touches something inside them that needs the attention of a counselor or therapist. Your instructor can help you recognize those instances and locate appropriate resources to deal with them.

The most important component of support is listening. It sounds simple, but think about how rare it is that someone really wants to listen to you, especially when you are struggling. Attention wanders, small-talk intrudes, unwanted advice is given. It is a wonderful experience just to have someone listen quietly, attentively, and empathically to whatever you want to say. It is also a wonderful gift to give another person.

> *Knowing that others are feeling similar feelings doesn't make those feelings go away, but it does make me feel better about having them. I feel more comfortable now opening up to my classmates and feel better equipped in encouraging them.*
>
> STUDENT JOURNAL ENTRY

The importance of the seminar to the quality of the internship cannot be overstated. A workshop given at the National Society for Experiential Education described the seminar as "a keystone to learning" (Hesser & King, 1995) because the quality of learning is enhanced when interns come together as a community of learners. It is further enhanced when the seminar is built around a conceptual framework such as the one we will offer you.

THE CONCEPTS UNDERLYING THIS BOOK

Experiential Education

An internship, like other kinds of field instruction, is a form of experiential education. While this approach to learning may not be well understood in many places on your campus, it comes out of a long theoretical and practical tradition, and we thought you might want to know something about it. Experiential learning has philosophical roots dating back to the guild and apprenticeship systems of medieval times through the Industrial Revolution. Toward the end of the 19th century, professional schools required direct and practical experiences as integral components of the academic programs, (such as medical schools and hospital internships, law schools and moot courts and clerkships, normal schools and practice teaching, forestry/agriculture and field work) (Chickering, 1977). Perhaps the best known proponent of experiential education was the educational philosopher John Dewey (1916/1944, 1933, 1938/1963, 1940). Dewey believed strongly that "an ounce of experience is better than a ton of theory simply because it is only in experience that any theory has vital and verifiable significance" (1916/44, p. 144). However, he was convinced that even though all real education comes through experience, not all experience is necessarily educative. This idea was reiterated by David Kolb (1984, 1985; Kolb & Fry, 1975), whose work was mentioned earlier and who emphasized along with Dewey the need for experience to be organized and processed in some way to facilitate learning. Dewey also felt strongly that the educational environment needs to actively stimulate the student's development, and it does so through genuine and resolvable problems or conflicts that the student must confront with active thinking in order to grow and learn through the experience.

Experiential education is based on the premise that for real learning to happen, students need to be active participants in the learning process rather than passive recipients of information given by a teacher. When learning is a passive process, teachers are the centers of energy who tell you the information that they think you need to know. But when learning is an active process, students are the centers of energy. The teacher's role is to guide or facilitate your learning by taking an interest in your work and coaching you through the experience (Garvin, 1991). As an active participant in the learning process, you play a central role in shaping the content, direction, and pace of your learning.

Interns tend not to be aware that experiential learning is considered a controversial form of education by some academics. Experiential education programs can be enormously expensive to administer; their academic scholarship is frequently questioned; and they can be a management nightmare for the institution (Borzak & Hursh, 1977; Chickering, 1977; Garvin, 1991; Hursh, 1981; B. K. Simon, 1972). Yet, field instruction has evolved into a disciplined, goal-focused form of education that has become the cornerstone of many professional as well as liberal arts programs of study. Perhaps experiential education continues to thrive in spite of its inherent problems because of what you and other students want from your education: opportunities for experiential, personal learning that has specific and unique meaning (B. K. Simon, 1972).

An internship is a particular form of experiential education, and it has some features that are perhaps unique. For one thing, in addition to knowledge and skill acquisition, the internship is also intended to promote self-understanding, self-discipline, and self-confidence (Borzak & Hursh, 1977; B. K. Simon, 1972). These are not extras or frills; they are right at the heart of the experience.

Predictable Stages

This particular book, while grounded in experiential education theory, is based on two "big ideas." The first is that interns go through predictable stages of development during the course of the internship. Over the years, as we have supervised interns, listened to their concerns and read their journals, talked with their site supervisors, and discussed internships with colleagues and students at other institutions, a predictable progression of concerns and challenges began to emerge. We have organized these concerns into stages, which are modeled after the work of Lacoursiere and Schutz (Lacoursiere, 1980; Schutz, 1967; Sweitzer & King, 1994). Understanding this progression of concerns will help you, your instructor, and your site supervisor predict and make sense of some of the things that may happen during the course of your placement and think in advance about how to respond. It will also help you view many of your thoughts, feelings, and reactions as normal, and even necessary. The experience then becomes a bit less mysterious, and for some people, that makes it more comfortable. For example, if you are feeling excited but also pretty anxious as you begin the placement, you may wonder whether that anxiety is a sign of trouble or where it may have come from. Knowing that it is a common and predictable experience will help you stop worrying about the fact that you are anxious and let you direct your energy toward moving through that anxiety. Here is what one of our students had to say:

> Now that I know I am not the only one that is concerned with these feelings I am better able to share them with others without feeling embarrassed.

Self-Understanding

The second major idea behind this book is that to make sense of your internship, you need to understand more than a stage theory. You need to understand yourself. No two students have the same internship experience, even if they are working at the same agency. That is because any internship experience is the result of a complex interaction between the individuals and groups that make up the placement site and each individual intern. You are a unique individual, and that uniqueness influences both how people react to you and how you react to people and situations. You view the world through a set of lenses that are yours alone. Therefore, each of you will go through these stages at your own pace and in your own way. Events that trouble you may not trouble your peers, and vice versa. Some of you will be very visible and dramatic in both your trials and your tribulations. Others will experience changes more subtly and express them more quietly.

Finally, no one experiences an internship in a vacuum. You have a life outside the placement (although it may not seem like it sometimes), and your network of family, friends, and academic and professional obligations will shape your experience in a powerful way. We want to help you think about yourself throughout your internship in ways that we believe will lead you to important insights and to a smoother journey on your path to personal and professional development.

In summary, then, the stages of an internship will help you understand internships in general and some of the things that are liable to happen during your internship. Understanding yourself will help you recognize the particular style in which you will experience the internship. Putting both of those pieces of knowledge together will give you a powerful tool to understand what is happening to you, to meet and deal successfully with the challenges you face, and to take a proactive stance in making your internship the most rewarding experience it can be.

OVERVIEW OF THE TEXT

This book is organized into four sections. In Section One, we present the conceptual framework that underlies the book. Chapters 2 and 3 focus on self-understanding, and we ask you to look at and think about yourselves through a variety of lenses. Chapter 2 is especially theory-intensive. Many different approaches to self-understanding are discussed, and we have tried to be as user-friendly as possible in discussing the theories. Although it is perfectly possible to use and integrate all of the theories, we anticipate that some instructors and/or students may want to choose which of the theories and approaches they will emphasize, especially if some of the theories and ideas are new to their programs. Finally, Chapter 4 introduces you to the stages of an internship.

In Section Two, we deal with the issues and concerns associated with getting started in an internship. We call this the *anticipation* stage. Some of our colleagues and students have told us that they prefer to deal with some of these issues and concerns before the actual placement, and this section could also be used in that way. Chapter 5 discusses the anticipation stage itself, and Chapters 6 and 7 help you become oriented to clients, co-workers, supervisors, and the agency itself. These two chapters could easily be read out of order.

Section Three looks at the challenges that await you after the initial concerns have been resolved. In Chapter 8, we help you take stock of where you are and where you want to be. We also try to help you anticipate some of the more common problems that interns encounter at this point. The final portion of Chapter 8 deals with the *disillusionment* stage, which can be a disheartening stage for students, faculty, and site supervisors alike, but which we view as a normal and necessary part of the experience. In Chapter 9, we give you the tools to move from disillusionment to *confrontation* as you attempt to identify and resolve issues that are standing in the way of your continued progress. We emphasize that this process is a learning experience in and of itself as well as one that paves the way for future learning. There is a problem-solving model presented in Chapter 9, although a different model could also be used if there is one you or your instructor prefers, or to which you have already been exposed.

The final section of the book examines issues and concerns that are common in the latter stages of the experience: *competence* and *culmination*. Chapter 10 deals with several professional issues, and it is here that we deal briefly with ethical issues. Of course, ethical issues can arise throughout the internship, but it has been our experience that interns often do not notice them, regardless of whether and how often they are covered in class, until some other concerns have been resolved. We suggest some common ethical issues that arise in an internship and also provide a decision-making model. Students who have had a course on ethical issues, or who have encountered this theme in many of their courses, may already have a model for thinking about and resolving ethical dilemmas. However, the issues raised here are probably new to you, or at least you may be encountering them for the first time in a professional context. The last chapter of the book attempts to help you end the internship in a productive way. It covers ending well with clients, co-workers, and supervisors, and there is a final reflection on the experience.

For Further Reflection

1. The beginning of your internship is a good time to review your academic program's expectations of you, your supervisor, and your instructor during the internship. This is particularly important in terms of knowing what you can expect from others and what others, including the staff at the placement site, might expect from you. Take time now to locate any written documents from your program that specify these responsibilities. Make copies and keep them with your other internship paperwork.
2. Internship students often use a language of their own. Your supervisor or co-workers may appear puzzled when you use certain terms, even though they are commonly understood on your campus. We call this language *fieldspeak*. There is also *agency-speak*, which you may adopt without even thinking about it after a few days at the placement, but it will puzzle your seminar classmates or even your instructor. Review your program's definition of terms and compare them to the ones in this

chapter. Be ready to explain them to people at your placement site. Also keep track of agencyspeak as you go along and don't forget to clue in your instructor and classmates!

3. Seminar class can become an important part of the internship experience. Now is a good time to think about what you want from that class. For example, what are your major objectives for the time spent in seminar? What works best for you as a way of teaching and learning? What role do you see yourself having in class?
4. As you begin your journal, you need to decide whether it will be handwritten, tape recorded, typed, or computer generated. Then you will need to make sure you have the materials you need. It is also time to clarify the larger purposes of your journal, who will have access to it, and how you want to set it up to meet those needs.
5. Consider using your journal as a portfolio of your internship. What sorts of things might you include? If this is not appropriate in your particular case, you can always keep a portfolio that is separate from your journal.

For Further Exploration

Collison, B. B., & Garfield, N. J. (1990). *Careers in counseling and development.* Alexandria, VA: American Counseling Association.

Somewhat dated, but an excellent discussion and taxonomy of a variety of different careers, with information on training and employment opportunities for each one.

Russo, F. X., & Willis, G. (1986). *Human services in America.* Upper Saddle River, NJ: Prentice Hall.

A good introduction to the six major categories of human service delivery systems as well as an introduction to public policy in human services.

References

Borzak, L., & Hursh, B. A. (1977). Integrating the liberal arts and preprofessionalism through field experience: A process analysis. *Alternative Higher Education,* 2(1), 3–16.

Chickering, A. W. (1977). *Experience and learning: An introduction to experiential learning.* Rochelle, NY: Change Magazine Press.

National Crime Prevention Council. (1988). *Reaching out: School-based community service programs.* Washington, DC.

Dewey, J. (1944). *Democracy and education.* New York: Macmillan. (Original work published 1916)

Dewey, J. (1933). *How we think.* Lexington, MA: D.C. Heath.

Dewey, J. (1963). *Experience and education.* New York: Macmillan. (Original work published 1938)

Dewey, J. (1940). *Education today.* New York: Greenwood Press.

Garvin, D. A. (1991). Barriers and gateways to learning. In C. R. Christensen, D. A. Garvin, & A. Sweet (Eds.), *Education for judgment* (pp. 3–13). Boston: Harvard Business School Press.

Hesser, G., & King, M. A. (1995). *Internship seminar: A keystone to learning.* Workshop presented at the annual meeting of the National Society for Experiential Education, New Orleans.

Hursh, B. A. (1981). Learning through questioning in field programs. In L. Borzak (Ed.), *Field study: A sourcebook for experiential learning* (pp. 259–266). Beverly Hills, CA: Sage Publications.

King, M. A. (1995). *Toward rewarding field experiences: A guiding framework.* Unpublished manuscript, Fitchburg State College.

Kolb, D. A. (1984). *Experiential learning: Experience as the source of learning and development.* Upper Saddle River, NJ: Prentice Hall.

Kolb, D. A. (1985). *Learning style inventory.* Boston: McBer.

Kolb, D. A., & Fry, R. (1975). Toward an applied theory of experiential learning. In C. Cooper (Ed.), *Theories of group process* (pp. 33–57). New York: Wiley.

Lacoursiere, R. (1980). *The life cycle of groups: Group developmental stage theory.* New York: Human Sciences Press.

Royse, D., Dhooper, S. S., & Rompf, E. L. (1996). *Field instruction: A guide for social work students* (2nd ed.). New York: Longman.

Schutz, W. (1967). *Joy.* New York: Grove Press.

Simon, B. K. (1972). Field instruction as education for practice: Purposes and goals. In K. Wenzel (Ed.), *Undergraduate field instruction programs: Current issues and predictions* (pp. 63–79). New York: Council on Social Work Education.

Simon, E. (1989). Field practice survey results. In C. Tower (Ed.), *Field work in human services*: Council for Standards in Human Service Education (Monograph #6).

Sweitzer, H. F., & King, M. A. (1994). Stages of an internship: An organizing framework. *Human Service Education, 14*(1) 25–38.

Weinstein, G. (1981). Self science education. In J. Fried (Ed.), *New directions for student services: Education for student development* (pp. 73–78). San Francisco: Jossey-Bass.

Weinstein, G., Hardin, J., & Weinstein, M. (1975). *Education of the self: A trainers manual.* Amherst, MA: Mandella.

Williams, J. K. M. (1975). The practice seminar in social work education. In J. K. M. Williams (Ed.), *The dynamics of field instruction* (pp. 94–101). New York: Council for Social Work Education.

CHAPTER 2

Understanding Yourself

> *I have been in my placement for several weeks and have challenged my own philosophies many times. It frightens me to think that the very foundation on which I have based my life is being challenged by clients who I believed were going to be textbook cases. Not that I assumed that I was entering a vacuum, but I didn't think that my own beliefs could be shaken in such a short period of time. Maybe I am making no sense at all. Maybe I am trying to make too much sense.*
>
> STUDENT JOURNAL ENTRY

As you prepare to start your placement, what are some of the things you'd like to know? What knowledge, if you had it, would make you feel more prepared? When we have asked students these questions over the years, we have gotten all kinds of answers, including those focused on clients, co-workers, intervention techniques, and agency rules. Those are all good answers and important things to know. What we hear less of, and what is just as important, is that you need to know about yourself.

Robert Kegan, a developmental psychologist, believes that the most fundamental human activity is "meaning making" (Kegan, 1982, 1994). All of us are engaged, all the time, in a process of trying to understand what is happening to us; we struggle to make sense of the world in which we live and the people in it. Perhaps nowhere is this activity more important than in human service work. In your preparation for work as a human service professional, you have been encouraged to consider the complex factors that lead to human problems and their solutions. As soon as you enter your placement, that will stop being an abstract exercise; you won't be thinking about human problems in general anymore, but about the ones that confront you daily. When that happens, you must consider an additional factor in thinking about problems, and that is *you*. You must pay attention not only to the sense you are making of what you encounter but of how you are making sense. Doing so will make you a more effective intern, a more effective practitioner in the future, and more sensitive to your individual journey through the stages of an internship.

Suppose, for example, you are confronted with the following situations at your internship:

- Your supervisor, in correcting your work, makes some insensitive statements.
- One of your co-workers is too aggressive in advancing some ideas.
- One of your clients is backsliding.

Now you must make sense of these events. What factors will you consider? What is the problem and who or what needs to change? The very words we just used to describe the situations imply that the problems, and hence the solutions, lie in other people. However, you are part of the problem too, and not just in what you may have done to cause another person's response. For example, you think your co-worker is being too aggressive. Too aggressive for whom? Some people would not find that behavior too aggressive or even describe it as aggressive at all. Why did you? What needs to happen now? Does your co-worker need to tone it down or do you need to learn to be more tolerant?

Self-understanding plays an important role in helping human service workers fulfill their responsibilities effectively and responsibly. As a human service intern, regardless of your specific duties, you are usually trying to form relationships with clients. In the context of those helping relationships, you attempt to respond, or help clients respond, to a variety of human problems. Although you may never have thought of it this way, in fulfilling these functions, you occupy a position of power and influence over clients (Brammer, 1985). Not only do you form opinions and make judgments, but you also make those opinions known to others in counseling sessions, reports, and staff meetings.

Furthermore, you may be asked to recommend for or against various kinds of intervention. Sometimes the placement you work for controls vital resources, in some cases as basic as food and shelter. Increased understanding of the feelings, beliefs, tendencies, and style that make up your system of meaning making will help you be more effective in all your functions. It will also help you be sure you are using your power for the welfare of your clients.

A number of authors have commented on the role of self-understanding in forming effective relationships (Brill, 1998; M. S. Corey & Corey, 1998; Schram & Mandell, 1997). If you can clarify and discuss your feelings and personal patterns, you can serve as examples for clients who are struggling to do the same. Furthermore, if you have confronted and modified aspects of yourself that you didn't like, you are more likely to communicate a belief in the capacity for change (M. S. Corey & Corey, 1998). Finally, awareness of ways in which your experiences are similar to those of clients helps establish empathy and trust.

However, you will also deal with many people whose experiences and values are very different from your own. In those cases, self-understanding will help you avoid three major pitfalls: (a) projection, (b) professional myopia, and (c) the tendency to regard difference as deviance (Sweitzer & Jones, 1990). At the heart of combating these tendencies is the ability to see your own reactions and views as only one among many possibilities. For this reason, understanding some of the sources of your personal characteristics and responses is critical. Understanding how your family, for example, influenced your beliefs and personality, coupled with knowledge of the diversity in family patterns and dynamics, will let you be more objective in your assessment and opinion of others.

Projection, as we are using the term, refers to the tendency to believe that you see in others feelings and beliefs that are actually your own. This tendency can affect your ability to understand, accept, and empathize with another person. If a client brings up an issue with which you are uncomfortable in your own life, you may avoid the subject or decide that the client is being difficult (G. Corey, Corey, & Callanan, 1998). If a you are angry at a client and unaware of that anger, you may decide that the client is angry at the world; if you have trouble with assertiveness, you may react negatively to an assertive client, supervisor, or co-worker. Projection can also lead to poor decisions about client needs. You may fall into the trap of inadvertently using clients to meet your own needs (G. Corey, et al., 1998). For example, if you have trouble with assertiveness, you may push clients to be assertive in part because it fulfills your need vicariously. Increased self-awareness will help you make more conscious choices in all these situations.

Self-understanding also helps you avoid something called *professional myopia*. Because everyone has their own system of meaning making, not only will two people see the same thing differently, but they will often see two different things altogether. If you are not aware that your point of view is only one of many, you may fail to consider other possibilities before making decisions about causes and solutions. For example, suppose you are an intern in a predominantly White high school and you hear a Black student complain of feeling out of place and uncomfortable, perhaps even unable to work to capacity. Part of the problem may be that certain features of the school itself, such as staffing patterns, menus, and recreational opportunities, are suited to middle-class White students. If you are not sensitive to these dynamics, you may describe the problem as one of adjustment or even cultural deprivation. Your efforts, then, will be aimed at helping the Black student feel better about a situation that may in fact be unfair or discriminatory.

When people are aware of differences between themselves and others, they sometimes confuse *difference* with *deviance*. They do not describe it as something different; they describe it as something wrong. You are not going to like every client, nor everyone you work with, but making the distinction between difference and deviance can help you accept and empathize with a wider variety of people. The tendency to assess difference as deviance can also lead to poor choices in using power and influence. If you do not see your perceptions as one of many possibilities, you may react to different perspectives by trying to change them. A trait assessed as a flaw in a client's personality, for example, may in fact be a cultural difference (Axelson, 1999). A family may choose to sacrifice some degree of their children's comfort and education to allow elderly grandparents to live in relative comfort. The family then becomes the target of a complaint by the school system. Their decision may reflect a cultural value about the relative responsibility families have to children and elders. If you do not see the decision in this light, you might condemn the family for their treatment of the children, thereby damaging your relationship with them. Furthermore, you might then choose to concentrate on changing the family, never considering the option of urging the community to accommodate a wider variety of family customs and structures.

Finally, the more you understand about yourself, the better you will be able to recognize the stages of the internship as they happen. It will not be enough to know, or suspect, that you are in Stage 1 or Stage 2. What does that mean for you? What strengths can you draw on and what personal traps should you avoid? To move through the stage you are in, are there aspects of yourself, as well as aspects of the internship, that you must confront and try to change? If you can discuss these issues in your journal and talk about them with peers, instructors, and supervisors, you will recognize the challenges sooner and move through them more smoothly.

It is important to realize that self-knowledge is not like other knowledge. Once you have learned a fact or a skill, you have it (unless you forget it or get rusty). But you are changing all the time, and your system of making meaning is changing with you. In addition, just as there is always something new to see and appreciate in a work of art, there is always more to know about yourself. So, self-understanding is a process to which you must become accustomed and committed. It takes work, but it will pay you big dividends.

Of course, self-understanding is an enormous area, and one that requires a continuous commitment. We have chosen several topics within the realm of self-understanding to concentrate on in this and the next chapter. Our goal is to introduce you to these areas, to give you some things to think about, and to provide you with resources to explore these areas further. We are trying to summarize some complex theoretical information, some of which may be new to you. You and your instructor will need to decide which of these areas merit your time and energy. That decision may be made for your class as a whole, or different class members may choose to pursue different areas. We hope this chapter is a resource for you throughout your internship.

VALUES

A value is an idea or way of being that you believe in strongly, something you hold dear and that is visible in your actions. In the previous section, we argued that self-examination is important; that is a value for us. You may believe strongly in taking care of your family, in serving your country in some way, or in the existence of a deity. You are probably very aware of some of your values, especially ones that have been challenged, debated, or highlighted in the media (e.g., values about abortion or euthanasia). Others, though, are so much a part of you and are shared by so many people in your life that they don't seem like values—they seem like truths. For example, both of us always placed a high value on punctuality and even assumed at one time that people who were habitually late were irresponsible. We have come to understand, however, that not all cultures share the same view of time. It is important that you clarify your values and give some thought to how you will respond when faced with people who do not share them.

Values permeate your life; we could never list all the important areas. However, here are some that may be especially important in your internship. Answers to these questions, especially those that describe how you believe things "should" be, are important clues to your values.

Sexuality How do you feel about homosexuality? Bisexuality? Heterosexuality? Teenage sex? Premarital sex? Monogamy? Extramarital sex? Various sexual practices?

Family How should a family be structured? Are single-parent families okay? Should one parent stay home with the children? Should there be a "boss"? If so, who should it be? How should decisions be reached? Should grandparents or other relatives live with the family?

Religion How important is religion to you? How do you feel about other religions? Would you ever do something that a leader of your religion said was prohibited?

Abortion How do you feel about abortion for yourself? For others? Are there circumstances under which you believe it is morally wrong? Morally justified? Should teens be able to choose on their own? Should the father have a say? Have you always held this position?

Euthanasia Should a person be able to choose to end his or her own life? Under what circumstances?

Self-disclosure What kind of things is it okay to tell someone you hardly know? For example, would you tell that person your financial problems? Your family difficulties? Your income? Which of these things would you tell a close friend? Are there things you would never tell anyone?

Honesty How do you define this word? Are there different kinds of lies? Is it ever okay to lie? If so, when and to whom?

Autonomy Do you think people should normally make their own choices and accept responsibility for their lives? Or do you tend to see people as more responsible for one another? How large a role do you think fate plays in a person's life?

Work How hard do you think a person should work? What do you think about people who only work enough to get by, and no more? What about people who seem to have no desire to find a better job or make more money? Those who always push themselves to work harder, no matter what? How about people who don't want to work at all?

Diversity How do you think people who are different from you ought to be treated?

Hygiene How often do you think a person should bathe? Should a person use deodorant? What would you think of a person who showed up at your office with dirty hands? A dirty face? Dirty feet? Dirty clothes?

Conformity How important is it to go along with the crowd, or not to go along with the crowd?

Time How important is it for you to be on time? That others are punctual?

Alcohol Is moderate use of alcohol okay? Is it ever okay to get drunk? How often? Do you think alcoholism is a disease?

Drugs Are there some illegal drugs that you think are okay to use in moderation? If so, is it also okay to sell them? What are your thoughts about the use or abuse of over-the-counter drugs or prescription medications?

These questions may help you clarify some of your values. Think of how you feel about them in your own life, but also think of how you feel about them in general. You might, for example, believe that you would not have an abortion, but that people should be able to choose that option for themselves. It is also important to ask yourself how strongly you feel about these values and whether you are open to changing them. You might also want to think about how you came to have these values. Did you choose them consciously, after careful thought? Perhaps you are not really sure where they came from and why you hold them. Finally, you will almost certainly encounter someone at your internship who does not share some of your values, and you should think about how you might respond.

REACTION PATTERNS

Reaction patterns are ways that you respond—your thoughts, feelings, and actions—to particular kinds of situations. Some patterns are helpful and work well. Others are distinctly unhelpful; they do not appear to get you what you want, and yet you repeat them in spite of yourself. One of the most frustrating, stressful experiences people can have is one in which they find themselves doing something they don't want to do or reacting in a way they don't want to react. Here are some examples:

> A friend has seen your in-class presentation and offers some constructive criticism. As the conversation goes on, you find that you are getting angrier and angrier and are having a hard time listening. You keep coming up, mentally or verbally, with defenses for every criticism, and you imagine telling your friend off.
>
> You are at a party and people are talking about some hot topic. You have something to say but can't seem to say it. Since others are very vocal, it is easy, although frustrating, for you to just sit there. Later, someone says exactly what you were going to say and everyone seems impressed.
>
> You are struggling with an intimate relationship. A good friend asks you how it's going with that person, and to your surprise, you hear yourself saying that things are fine.
>
> A friend calls late at night and asks that you to meet him right away. The matter does not seem like an emergency, but you leave your homework undone and go off to meet your friend.

These are not situations in which you later find, after much reflection, that you made a mistake. They are situations in which you know immediately afterwards, or even during the situation, that you are not responding the way you want to. In fact, almost as soon as it is over, you can think of several ways to handle the situation that would have been better. Think about situations like this that have happened to you. Jot a few of them down.

You may find that these are isolated incidents or that they occur with just one person. You may also find, however, that these responses are part of a pattern for you. You may find that in general you are defensive about criticism, unable to speak in groups, unable to say "no" even when you want to, or to ask for help. Gerald Weinstein (1981) describes these tendencies as dysfunctional patterns. The word *dysfunctional* has become very popular lately. Please note that, in this case, dysfunctional does not mean useless, nor does having a dysfunctional pattern make you a dysfunctional person. On the contrary, these patterns are clues to some important aspects of yourself. We both have them, and so does everyone else. The reason we are asking you to think about these patterns is that you may "bump into" them during your internship. We will talk more about what patterns you might encounter and what to do about them in Chapters 8 and 10.

See if you can identify any such dysfunctional patterns in your life, and try using the format suggested by Weinstein:

> Whenever I'm in a situation where __________, I usually experience feelings of __________. The things I tell myself are __________, and what I typically do is __________. Afterwards I feel __________. What I wish I could do instead is __________.

Here is an example to help you:

> Whenever I am in a situation where I feel angry at a friend, I usually experience feelings of anxiety and self-doubt. The things I tell myself are, "Take it easy. It's not that bad. There's probably a good explanation, and besides, you don't want to upset him." What I typically do is smile, joke, or protest very weakly. Afterwards I feel as if I let us both down. What I'd like to do is find a clear, respectful way to tell my friend what is upsetting me.

And here is one from an intern's journal:

> I am trying to build my self-esteem and confidence. It is hard for me to hold back my emotions when something upsets me, especially my fear of doing something wrong or failing. I appear to be happy and cheery, but it takes so much out of me that I become negative and sometimes grumpy. This gets in the way of me liking myself.

LEARNING STYLES

One of the most important things you can discover about yourself is how you learn best. You have probably heard the labels "fast learner" and "slow learner" applied to other students or even to you. Recent research in educational psychology has shown these terms to be a bit clumsy. What seems to be true is that there are different styles of learning and that some learning experiences require certain styles. If you are engaged in a learning experience that is not well suited to your style, it may be a struggle. Certainly, you can think of times when you learned something easily and other times when it was more difficult. Your internship is going to provide you with a variety of experiences to learn from, and some will be easier than others.

If you are involved in an experience that is not well matched to your style, you can still learn, but something needs to give. You may be able to take steps to change or augment the learning experience so that it is better suited to your strengths. For example, if you prefer abstract modes but are given a great deal of concrete experience, you can ask for some readings or find some on your own. On the other hand, you may need to try to stretch your learning repertoire and strengthen a style that does not come easily to you.

Knowledge about learning styles can also help you be more effective with clients as you help them learn to do new things such as change their behavior, locate resources, or improve their parenting skills. If you can learn to adapt interventions to a variety of learning needs, you will be more successful in this attempt. If you understand diversity in learning styles, you are less likely to view those different from your own as deviant or deficient.

You may have studied learning style in some of your classes; there are a number of theories to draw upon. The variety of these theories can be exciting, but it can also be confusing. Claxton and Murrell (1987) have pointed out that many theorists use the term *learning style* to describe qualitatively different aspects of learning, thus making a single definition difficult. It would not be accurate, then, to speak of a student's learning style unless referenced to a specific theory, because learning style is not a single, static trait, but a combination of many characteristics and tendencies. We have chosen two areas in learning style to call to your attention. The first involves the way you take in and make sense of information; it draws on the work of David Kolb. The second concerns the way you relate to knowledge in general; it draws on the work of several psychologists and educators.

Kolb's Theory

David Kolb (1984) originally set forth a cycle of four phases that people should go through to benefit from experiential learning, as illustrated in Figure 2.1. The first phase is concrete experience (CE); students have a specific experience in the classroom, at home, in a field placement, or in some other context. They then reflect on that experience from a variety of perspectives (reflective observation, or RO). During the abstract conceptualization (AC) phase, they try to form generalizations or principles based on their experience and reflection. Finally, they test that theory or idea in a new situation (active experimentation, or AE) and the cycle begins again, since this is another concrete experience.

You may recognize this cycle from your internship. For example, suppose you observe a client in an argument with a staff member. You could then draw on several theories or ideas you have learned to try to understand what happened, or you might seek out some new information from staff, books, or articles. You then begin to form your own ideas about what happened and why, and you might use this knowledge to guide your own interactions with that client. Once you do that, the interaction is itself a new concrete experience, and the cycle begins again.

Kolb's four phases can also be thought of as essential abilities that all students need to develop (Sugarman, 1985). You will most likely be stronger in some areas than others, and different learning activities call on and develop various abilities. The concrete experience phase requires you to involve yourself fully, openly, and without bias in an experience, avoiding the rush to analyze and interpret. Active listening, which you may

FIGURE 2.1
Kolb's Experiential Learning Cycle

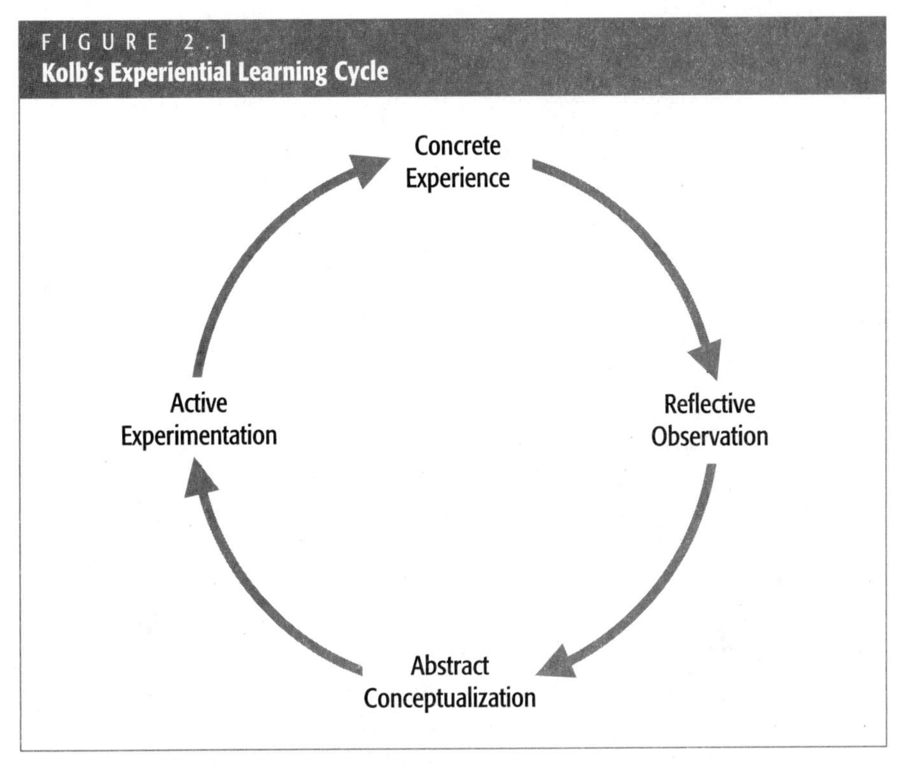

Source: Kolb & Fry (1975).

have studied, is a perfect example of these abilities. If a client or colleague is talking to you, it is difficult to really listen if you are busy thinking about what the hidden meaning might be or what you are going to say when they are finished. Reflective observation requires the ability to consider an experience from a variety of perspectives. You must be able to slow down and apply several different perspectives (which may contradict one another) before forming conclusions. In your internship, you might be asked to analyze a case or a problem using a variety of theoretical frameworks. In abstract conceptualization you begin to form theories or principles of your own. The ability to integrate is important here, as well as the ability to synthesize and generalize. In the active experimentation phase, you test your ideas, which means that you must be willing to take risks and make mistakes.

Although the experiential learning cycle begins with concrete experience and continues in order, Kolb also points out that there are two fundamental dynamics involved in learning: taking in information and processing it in some way. Each of these dynamics can be approached in a concrete or abstract way. The concrete experience phase of Kolb's cycle is a way to take in information in a very concrete way—through experience. However, it is also possible to take in information in a more general, abstract way, such as through reading. This is abstract conceptualization. When information is processed in a concrete way, it is through experimentation (AE); when it is processed abstractly, it is through reflection (RO).

As you read, you will no doubt recognize your own strengths and weaknesses. It is certainly useful to know your tendencies, and there are both formal and informal techniques for discovering them (Kolb, 1985). However, there is evidence that exposure to and competence in a range of learning styles help you become a more flexible and complex thinker (Kolb, 1984; Stewart, 1990). So, for example, if you are assigned a number of readings about various forms or procedures at your agency, that will be easier to learn if you prefer an abstract mode of taking in information. If you do not, you might learn the same information more easily by practicing on actual (concrete) cases.

Most internships ask you to be active very soon and to learn from experience. This can be of great value, and some students who have struggled in more abstract classes feel much more competent and at home in the internship, where they are expected to learn by doing. However, if you tend to learn best in an abstract manner, you may feel a bit lost. When these mismatches occur, we encourage you to see what you can do to give yourself the kind of learning experiences you need—seek out some reading, for example, or ask for some more experience. However, you should also view the internship as a chance to stretch your learning style repertoire by becoming more adept at styles that you now find difficult.

Separate and Connected Knowing

Some recent work in psychology has focused on the elaboration of two basic approaches to the world: separation and connection (Gilligan, 1982; Gilligan, Ward, & Taylor, 1988; Lyons, 1983). Separation is an approach to the world that emphasizes autonomy and abstract principles. Connection, on the other hand, emphasizes relationships and the importance of context. Much of this work was originally undertaken to fill a gap in developmental psychology caused by the use of predominantly male samples in psychological research and observation. By focusing on the experience of women, scholars were able to uncover important aspects of human development. Because of these origins, this work is often referred to as work on *women's development.* However, it appears that both orientations are found in men and women, although it also appears that the connection orientation is more often found in women.

The ideas of separation and connection were applied to learning in the book *Women's Ways of Knowing* (Belenky, Clinchy, Goldberger, & Tarule, 1986). Separate and connected knowing are terms used to describe opposite ends of a spectrum. Most people show some evidence of both styles, and most also lean distinctly in one direction or the other. Each way of knowing has strengths and limitations. As you read the descriptions of these two styles, see whether you can recognize yourself.[1]

Separate knowers try to understand through analysis. They apply the principles learned in various disciplines. The scientific method, which you have probably studied, is an example of such a set of principles that is used to try to uncover scientific truths.

[1]*Women's Ways of Knowing* is a complex and interesting book. It identifies five epistemological positions taken by women (and presumably by some men as well). The distinction between separate and connected knowing is made within one of those positions, called *procedural knowledge,* which is a position commonly found among college students. We urge you to explore this work further.

You may also have learned methods for analyzing a poem, a play, or a painting. To use this approach, you must be familiar with the specifics of what you are trying to understand, but not overly focused on them. Consider the metaphor "they cannot see the forest for the trees." You certainly cannot see a forest if you are not paying at least some attention to trees. However, if you focus on each individual tree, you will never see the patterns of tree growth in the forest. You will see the unique features of each tree, but you may lack a set of terms with which to compare them to one another. You also could see the beauty in each tree, but be unable to say which is a better tree. You may have been similarly mystified about dog, cat, or horse shows. Although the animals look different, each looks truly beautiful; how do the judges discriminate among them? Principles of analysis let us compare and evaluate. Two works of art may use different media, come from different times in history, and have been created in very different emotional contexts, but they can be compared and judged as works of art. Separate knowers, then, tend to separate themselves from too many specific details and look instead to organize experience according to abstract principles.

Turning now to the understanding of people, disciplines such as psychology and sociology offer separate knowers different lenses through which to make sense of human behavior. Each theory within these disciplines offers a different lens as well. Theories of human development (such as those of Piaget, Gilligan, and Erikson) as well as psychological theories (such as cognitive behaviorism, Gestalt, and transactional analysis) offer distinct ways of examining a person's behavior and reactions. A separate knower tries to understand a person by applying such theoretical frameworks.

Separate knowing is often used to criticize ideas or positions. Separate knowers need to be convinced of the truth of an idea or the authenticity of a person, and they expect to be convinced using agreed upon principles of analysis and persuasion.

Connected knowers, on the other hand, believe that truth is personal, specific, and located in experience. They attempt to understand not by analyzing but by empathizing. Rather than apply abstract principles, they immerse themselves in specifics. In examining a poem, for example, a connected knower is more likely to try to imagine what the poet was feeling, or intending, than to break the poem down into components for purposes of comparison. Separate knowers, by contrast, may miss unique aspects of a situation or work of art that do not fit the model they are using. It's not that they see these aspects and choose to ignore them; they literally do not see them. A connected knower may have a hard time applying a theory or abstract principle because each situation, each work of art, is sought after and seen in its own terms.

If separate knowing is used for criticism, connected knowing is used for developing trust and empathy. Connected knowers approach an idea or person with the belief that there is something of merit there, and they try to find it.

In attempting to understand people, connected knowers try to share the experience of the person they are talking with or thinking about and understand what created that experience. They are less drawn to theories as a way of making sense of a person's experience. Separate knowers, in applying psychological or counseling theories, may focus more on what they are supposed to say, or on how the person's statements can be interpreted from the particular theoretical paradigm, than on actually listening. Table 2.1 summarizes the key differences between these two ways of knowing and approaching the world.

TABLE 2.1
Two Ways of Knowing

	Separate	**Connected**
Mode of Understanding	Analysis Abstract Principles	Emphathy Immersion
Understanding People	Phychological and Sociological Theory	Experiential Logic of Each Individual
Relationship to Ideas	Skeptical Critical	Look for Merits

Perhaps you recognize some of your tendencies from these descriptions. Human services is an arena where connected knowing is valued, as opposed to being a liability, as it may be in many learning situations (Belenky et al., 1986). You will need to use both separate and connected approaches in your internship. What is important is to know where your natural strengths lie, what to use when, and where you will need to stretch and grow.

Many of your courses up to now probably emphasized the learning of principles and theories. In your internship, though, you are going to be immersed in experience and in specific situations. The theories you know will be only as good as your ability to use them in service to clients and the organization. We have known many students who struggle some with theory but are outstanding in the field, and vice versa.

Theoretical knowledge is important in an internship. When you counsel a client, conduct an interview, plan a change strategy, or organize a project, theories help you know how to proceed. Although it may seem that understanding a theory is easier for a separate knower, both separate and connected knowers can learn theory. In fact, connected knowers' tendency to want to "get into the head" of the theoretician may help them achieve a richer and fuller understanding of the theory. However, when it comes to using that theory to analyze a specific person or event, they may struggle.

In working with clients, a theory may help you understand, but it will not necessarily help you connect. After all, you are not interacting with a developmental stage or a psychiatric syndrome; you are interacting with a person. Most people wish to be seen as unique, not reduced to a stage or a diagnosis. The more you discover the experiential logic of that particular person (Belenky et al., 1986), the less strange that person's behavior and reactions will seem, and the better you will be able to communicate empathy and acceptance. On the other hand, if you use only the client's frame of reference, you lose the opportunity that theories provide to get a fresh look and to step away from repetitive, destructive patterns and rationalizations.

In your seminar class or support teams, there is a time and place for both separate and connected approaches. If you or someone else is struggling with a problem at your placement, a theoretical analysis can help you look at it in a new way. At the same time, a connected approach helps ensure that each person's unique experience is attended to and honored. Understanding the difference between separate and connected knowing may also help you understand your supervisor and co-workers better. Their expectations of and reactions to you may be based in part on their approach as separate or connected knowers. We will return to these themes later in the book.

FAMILY PATTERNS

For many people, their family is the group with which they spend the most time until they form their own family or leave home for some other reason. Your experience in your family is a powerful influence on who you are. Each family has its own way of doing things. Often, we are so accustomed to the way our family is that we assume that everyone's family is that way. For example, while both of us come from families where eating dinner together was very important, they differed considerably in how that time was spent. For one of us, conversation at the table was quiet and polite, and interruptions were frowned upon. For the other, the dinner table was lively, and often loud, with many conversations taking place at the same time. Imagine the shock for one of us visiting the other's family for dinner or for either of us visiting a family whose members came and went from the table or ate in different parts of the house.

We are going to encourage you to think about two common features of family life: rules and roles. Although very few families have a list of rules posted on the wall, they all have unwritten rules that tell everyone what they can and cannot do. As a child, you usually learn the rules by breaking them; someone reprimands or disciplines you, and you gradually figure out what is and is not acceptable. Sometimes families have rules that they are not even aware of. For example, one of us once listened to a family over the course of a weekend as they talked about all their relatives, living and deceased. There was one who was never mentioned, though. The family was quite surprised when this observation was shared, but upon reflection agreed that they almost never talk about that person. In observing this informal tradition, they were obeying a rule, even though they would never have called it that.

Here are some examples of family rules. Use this list to stimulate your thinking about the rules in your family:

Don't talk about sex.

Keep family business in the family.

Guests are always welcome for dinner—you don't have to ask.

Never question a decision made by Mom.

Mom and Dad need 15 minutes to relax after work before you ask them anything.

If you need extra money, ask on a Friday.

No one is entitled to privacy, except in the bathroom.

Family roles tell you who in the family performs which functions. Some of them are pretty formal. One parent, for example, may pay the bills, do the cooking, or handle the discipline. There are other kinds of roles, though. There might be a family jokester, who is counted on to make people laugh, or the mediator, who tries to help settle conflict between family members. Some children get the role of the "good" child, and their mistakes are often overlooked or treated lightly, whereas others get the role of "bad" children and are treated more strictly. Here are some questions to help you think about roles in your family:

Who has the final say in an argument or dispute?

Which child is the smart one? The talented one? The athletic one?

Who can you count on to help you out of a jam?

Which child gets the most leeway from Mom? From Dad?

In your internship, you will undoubtedly meet people whose family patterns are different from your own. That can be pretty confusing as you try to make sense of their behavior and reactions. Understanding the sources of some of your behavior and feelings will, again, help you look more thoroughly and empathically at others. Also, if you are not careful, you may carry a rule or a role from your family into your internship that is not helpful in that context. For example, your role as the jokester in your family, while it may serve to ease tension at home, may be annoying in a staff meeting, especially one where there is tension. Similarly, your role as the comforter may make it difficult for you to let a client struggle through his or her own problems.

PSYCHOSOCIAL IDENTITY

You have probably had a course in developmental psychology in which you had some exposure to the ideas of Erik Erikson. In fact, you may have had a lot of exposure to Erikson, and are now thinking to yourself, "Erikson? Again?" Well, yes, but with a different twist. Although some of Erikson's ideas have been questioned and criticized since he first advanced them, some of the basic tenets of his work are valuable tools for self-understanding (Sweitzer, 1993). So, we are going to review those tenets and discuss four of the stages and how they may have an impact on you and your internship.

According to Erikson, the interaction between person and social environment generates a series of psychosocial issues for the developing person. These issues can be thought of as questions that are constantly on the person's mind. The question for the school-age child is, for example: Can I succeed? These questions are not general ones; they are asked and answered every day. Thus, the question for the school-age child is not whether he or she can succeed in general; it is whether he or she can succeed in this endeavor, right now.

These questions are experienced sequentially as the person progresses through the life span. However, an important tenet of Erikson's theory is that growth is epigenetic; the issues are visible, albeit in different form, before and after the stage in which they become central. Each issue, though, has a time in the life span during which it is particularly important (Erikson, 1982). So, for example, your struggles with trust had an impact on you as a small child, but you may also be able to see how you struggle with trust as an adult (hence the relevance to your internship).

As developing people struggle with these questions, they experience an interplay of *polarities*, which are the opposite extremes of possible answers. To continue the exam-

FIGURE 2.2
The Dynamics of Erikson's Psychosocial Stages

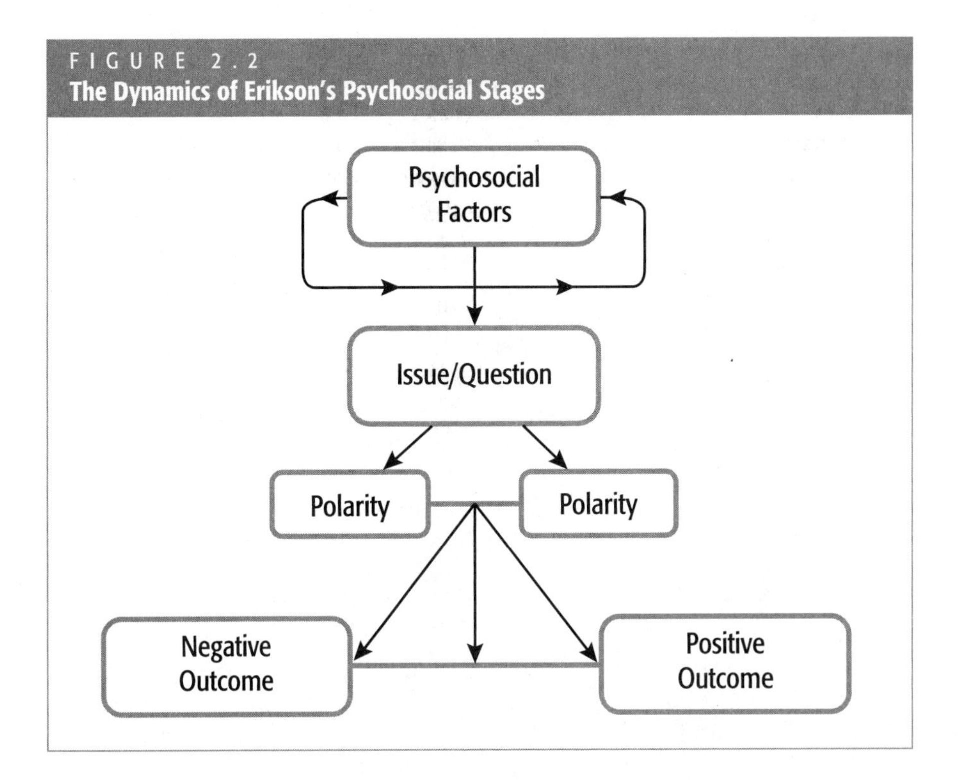

ple, when school-age children attempt something they either succeed or fail; success and failure are the polarities of this stage. Everyone will experience both polarities; not only is it impossible to experience only one, but both are important in determining the outcome of the psychosocial stage.

The experience of the interplay of polarities will eventually lead the developing person to what Erikson calls a *sense*. These senses are the outcomes of the psychosocial stage and can be placed on a continuum. These continua will be familiar to most of you because they are the names of Erikson's stages: trust vs. mistrust, autonomy vs. shame and doubt, initiative vs. guilt, industry vs. inferiority, identity vs. role confusion, intimacy vs. isolation, generativity vs. stagnation, and integrity vs. despair. A person emerges from a stage somewhere on the continuum between the two senses. These basic dynamics are represented in Figure 2.2.

However, there is no simple correspondence between polarities and outcomes. Erikson points out, for example, that there is no absolute correspondence between the quantity of love and attention the child receives and the degree to which he or she emerges with a sense of trust (Erikson, 1963). This principle continues throughout the stages; unlimited freedom does not necessarily produce autonomy, plentiful achievement does not always lead to industry, and so on.

We are going to encourage you to think about the issues involved in the first four stages. We believe that the questions raised during these stages are important components of who you are and deal with issues that will be significant for you as an intern.

In order to facilitate self-understanding, we will use Erikson's ideas in a slightly different way than he originally did.

Trust vs. Mistrust

This is the stage associated with infancy, with the individual infant is physically unable to meet its basic survival needs. The child, essentially helpless, interacts with a world of caregivers. The meeting of these needs, then, is the core issue; if they are not met, the child will die. The social environment of the infant is usually very limited. There are only a few people charged with the care of the child. If these people are unable to meet the child's needs, the child has nowhere else to turn, so the ability to depend on others is a critical issue. The central psychosocial question, then, is: Will my needs be met? This question is asked implicitly hundreds of times a day as the infant depends on the caregiver.

The polarities of the stage are simply need met or need not met. All children have experience with both polarities; they must have at least some needs met, but it is not possible to meet them all. Through the experience of both polarities, each child arrives at an outcome, which will be somewhere on a continuum between basic trust and basic mistrust.

Children who lean toward a sense of basic trust believe that even though their needs are sometimes not met, fundamentally they can count on the world to satisfy their needs. Note that they do not believe that their every need will be met; children with that belief are in for disappointment. They cannot develop a sense of trust unless they have the experience of denial. Children who lean toward a sense of basic mistrust, on the other hand, believe that even though their needs are sometimes met, fundamentally they cannot trust themselves or others to meet their needs. Furthermore, such children may feel badly about even having needs. That is a potentially crushing thought, since more than at any other time those basic needs are the core of the child.

For children with a sense of trust, a negative experience (one in which their needs are not met) does not disconfirm their fundamental view of themselves and the world. For children with a sense of mistrust, however, those events actually confirm the view that they cannot, and perhaps should not, have their needs met; they cannot trust their world.

To explore this part of yourself, try to find out about your experiences as a child. What is more important, though, is exploring this part of you as an adult. Were you a calm, easy baby? Were your parents and other caregivers able to provide for your needs most of the time? How did you respond when your needs weren't met? What is it like for you now when you really need something and don't get it? No one likes that experience, but some people find it devastating. You may also be a person who doesn't ask for much from others because you are afraid you won't get it.

In your internship, you may find yourself needing various things from your co-workers or supervisor. If they disappoint you, it is natural to be slightly upset, but a person with a shaky sense of trust may find these experiences unduly distressing. You may also find that, fearing your needs will not be met, you hide them, trying to appear confident and refusing to ask for help.

Autonomy vs. Shame and Doubt

For toddlers, the central issue has moved from needs to impulses. Children at this age are more mobile, agile, and have the beginning of language skills. They want to try out their new physical abilities, and this desire often results in impulsive behavior. Toddlers are experiencing the power of being able to do things themselves, of acting on impulses without the aid of an adult. The social environment is also changing, and there are increasing expectations for control. The child is expected, and told, to curb at least some impulses. As the impulsive child confronts these expectations, the interaction yields the central question: Can I act on this impulse?

The polarities of this stage, then, are independence (yes, you can) and control (no, you can't). Again, both polarities will be experienced. All children manage to get their way sometimes, and regardless of the strictness or permissiveness of an individual parent's philosophy, there will be times when the child must be told "no." Similarly, children's expression of feelings may be encouraged or thwarted. It is possible to let the child know that the feeling (the impulse) is okay, but the words and actions are not. Children's experiences with independence and control will influence where they end up on the continuum of outcomes. If a child is made to feel badly about even having the impulse or feeling, it will surely influence a more negative sense of self.

Children who end up toward the basic autonomy end of the continuum feel that even though they have to be controlled or curb their impulses sometimes, fundamentally their impulses are good, and they are capable of independent action. Instances where they are controlled (by others or by themselves) do not disconfirm their fundamental picture of themselves.

Children who lean toward the other end of the continuum, basic shame and doubt, feel that even though they can sometimes do what they want—complete an impulse—their impulses are suspect and shameful, and they must not surrender to them. They feel a basic sense of shame and doubt about their natural impulses.

This is a part of your life that you may have some memory of, and you may want to explore those memories or ask people who knew you then about their memories. Again, though, you should also focus on your current experience. Adults have impulses, too, and your experience in this Eriksonian stage may influence your feelings about impulses today. The issue is how you feel when you have an impulse, regardless of whether you act on it. Consider the example of a college student who has an exam the next day and is invited to go out with friends. She is tempted to go, and that temptation will bother her even if she refuses the invitation. Rather than being proud that she controlled the impulse, she feels guilty for having it. The impulse, not the control, has confirmed her fundamental view of herself. If she goes, she may drive her friends to distraction with her constant remarks about how she really should not be there. To surrender to impulse, regardless of how rarely, confirms her negative belief about herself.

If this is a vulnerable area, you may find yourself struggling with a particular kind of perfectionism at your internship. You may know how you should handle a situation, and you may even handle it that way, but you may have been tempted to behave otherwise. You may be tempted, for example, to be sarcastic, to yell at a client or co-worker, or to pretend you didn't see something. Both of us have supervised interns who become upset with themselves not for what they did, but for what they almost did.

Initiative vs. Guilt

During the preschool years, as the child continues to develop physically and cognitively, gratification of impulses ceases to be the main concern. It is replaced by the desire to plan and complete something, to explore, and to experiment. Erikson calls this desire the need to take initiative. Again, however, the child operates in a social environment that sometimes needs to limit experimentation and questioning. A child's desires for experimenting may be annoying, inconvenient, or even dangerous, as when the 4-year-old sets off alone to ride a bicycle. Certain lines of questioning may be inappropriate or concern matters considered private. So another conflict emerges, and the major question is: Can I finish what I start?

As has been true of questions in other stages, this question is asked many times every day. Eventually, the child will develop a general sense about this question and be somewhere along the continuum of outcomes for this stage. But for now, each time the question is asked it can have a different answer.

The polarities for this stage are initiative completed or initiative blocked. If children choose or are given manageable tasks and are allowed to complete them even if the result is not completely successful, the initiative is completed. If they take on too much or are not allowed to carry an idea forward, the initiative is blocked. Of course, each child experiences both of these polarities, as has been the case at every stage. Furthermore, the absolute amount of blocked initiative is probably not as important in determining the outcome of this stage as the way in which it is blocked. As is the case with impulses, children need to believe that their desire (in this case, the desire to explore, question, and experiment) is important and valuable, even if they are not allowed to complete a particular experiment or initiative.

From children's experiences with both polarities, they gradually settle into a place on the continuum of outcomes. At one end is basic initiative. Children with this sense believe that even though they sometimes may not be able or allowed to complete what they start, fundamentally they are capable of doing so and enjoy the inquisitive, adventuresome part of themselves. At the negative end of the continuum is a sense of guilt, which is felt by children who believe that even though they sometimes do complete what they start, their desire to explore and experiment is not a positive thing, and fundamentally they or the world will find a way to block these initiatives. The thwarting of an initiative, no matter how infrequent, confirms the fundamental belief these children have about themselves and reinforces feelings of guilt.

To explore the way this stage is present in your adult life, think about times when you completed a project or idea and times when completion was thwarted. Recall the way in which blocking happened and how you felt about it. Most people have no trouble recalling at least one or two significant instances. The absolute accuracy of the recollection is not important; the feeling that has stayed with you is the key. Then ask yourself how you feel now when you have an idea, make a plan or need to take charge of something. Does it feel like an adventure or more like a minefield? Think about your current reaction to being told "no" or having your desires frustrated in some way.

This psychosocial issue could arise at your internship when you have an idea for a project or a special approach with a client. A tenuous sense of initiative may discourage you from going forward and seeing it through. It could also arise when you are given a new or large assignment by your supervisor. You might find yourself terrified or automatically listing reasons why you cannot do it.

Industry vs. Inferiority

School-age children experience a significant change in their cognitive abilities as they become increasingly capable of logical thinking (Kegan, 1982). Their physical strength and coordination also increase dramatically, and of course, they want to test out these new abilities. They want to do more significant, adultlike tasks, and they have the capacity to do so. Changes in cognitive and social development also make this a time during which children begin to compare their performance to that of their peers. These desires to achieve, perform serious tasks, and compare favorably with peers often mirror the expectations of their social environment. Parents often begin to expect a sharing of household or child-care tasks and increasing achievement in school. In school, children are expected to learn, and that learning is often measured by some sort of performance such as a test, an athletic contest, or a recital.

The basic issue generated by the interaction of the child and the social environment is one of competence. The question for the child is: Can I succeed? The polarities are success and failure, and it is expected, and important, that each child experience both. Children with a sense of industry believe that even though they fail sometimes, they are basically competent individuals. Children with a sense of inferiority, on the other hand, feel basically incompetent, even though they may succeed sometimes (or even a lot).

The absolute amount of success and failure is probably less critical in determining the outcome of this stage than the reaction to success and failure. Certainly, all of us have known children (and adults) who have great confidence and resilience, even in the face of repeated failure. Similarly, there are children who seem anxious and unsure of themselves in spite of being very successful in school. For some children, any failure confirms their basic sense of themselves, regardless of how many successes precede it. Children who are praised only when they succeed, regardless of how much or little effort was required, may come to believe that success equals competence. Since no one can succeed all the time, their sense of competence may be tenuous, regardless of how much success they experience; they have a sense of basic inferiority. On the other hand, children who are praised for their efforts rather than the results, and who are encouraged to view failure as a learning experience and a natural occurrence, may develop a much more unshakable sense of competence. These children emerge with a sense of basic industry; they feel competent even in the face of failure. Children with a sense of basic inferiority, however, feel incompetent even in the face of success.

If you want to explore this stage of your life, try to recall significant experiences with success and failure. Remember how you reacted to these experiences and how those around you reacted. Then consider your current reactions to success and failure. Fear of failure or a feeling of basic incompetence can contribute to a number of dysfunctional patterns, including procrastination, test anxiety, and the aforementioned inability to participate in class. Fear of success, on the other hand, also contributes to a number of patterns, including ambivalence and anxiety about attaining your goals

TABLE 2.2
Psychosocial Challenges

Age Range	Individual	Social Environment	Issue	Polarities	Outcomes
Infancy	Cannot Meet Basic Needs	Limited to a Few People	Meeting Needs	Need Met Need Unmet	Trust vs. Mistrust
2–3 Years	Try Out New Abilities	Expects More Control	Acting on Impulses	Independence Control	Autonomy vs. Shame/Doubt
4–6 Years	Purposeful Action	Limits on Experimentation	Finishing What Is Started	Completed Blocked	Initiative vs. Guilt
Primary Grades	Accomplish Serious Tasks	Achievement Performance	Competence	Success Failure	Industry vs. Inferiority

or being less successful than your abilities would predict (Krueger, 1984; Schenkel, 1984).

The internship offers countless opportunities for success and failure, and you will experience both. An interaction with a client, an attempt to defuse an argument, a group activity, a phone conversation, and more could all go awry. No one likes when such occurrences happen, but individuals with a sense of inferiority may find them especially troublesome and difficult to overcome. You may also see clues to your position on this continuum in your reaction to criticism or evaluation. Table 2.2 summarizes the dynamics and issues found in each of these four stages.

YOUR CULTURAL IDENTITY

Do you see yourself as a prejudiced person? Do you hold any stereotypes about people of a different gender, race, ethnicity, sexual orientation, or religion? If you are like most of our students, you will answer these questions "no" or "not anymore." If you have qualified for an internship or community service project, you have probably been exposed to the core values of tolerance, pluralism, and respect for individuals that are at the heart of the helping professions. And naturally, you want to see yourself as a person who adheres to those values. If a part of you suspects that you do have some lingering prejudices, you may be keeping them hidden for fear that your peers or your professors will think badly of you or block your progress in your program.

Here is what we believe and invite you to consider. Everyone carries some stereotypes and prejudices, including us. If you work at it, you can learn to see your prejudices and make progress in overcoming them. However, at the same time, you will surely discover other, more subtle ones. The first step, though, toward being a nonprejudiced person is to confront and accept the prejudices you have. If you hide or deny them because you are ashamed or think that a good person would never have any, they will never change.

Try this exercise. Pick a subgroup that has been the target of some discrimination and of which you are not a member. Depending on your own subgroup membership, some examples are Blacks, Muslims, lesbians, blue-collar workers, Native Americans, and many others. Now, think about all the stereotypes you know about that group. Remember, we are not asking which ones you believe. Just write down as many as you can think of. You might want to get together with a classmate or two and expand your list. Now, look at your list. Where did the stereotypes come from? If you don't believe them, how did you come to "know" them?

Most students have little trouble coming up with a substantial list. The reason stereotypes are so easy to recall is that they are literally all around us. Think about the group you picked. How many members of this group do you actually know? In some cases the answer will be few or none. So where do your impressions of them come from? Some, of course, come from family and friends. Another part of the answer, though, is the way members of these groups are depicted in the media. Usually, members of various subgroups are portrayed in very limited, stereotyped roles. Think, too, about the books you read in school. How many members of these groups were in those books? How were they portrayed? The point here is that we have all "learned" harmful stereotypes about other groups (and even about our own).

Naomi Brill, in her excellent book, *Working With People* (1998), cautions human service professionals to be sensitive to their use of "the paranoid 'they.'" If you find yourself thinking that "they" are too aggressive, too concerned with money, or too clannish, that is a key to a stereotype or prejudice that you hold. We encourage you to explore and admit your prejudices and recommend some resources for doing so at the end of this chapter. Once you have identified some prejudices, you can work on changing them. For example, through an experiential exercise like the one we just encouraged you to do, Fred once discovered that he held some stereotypes about Hispanic men:

> I had to admit that I felt uncomfortable in groups of Hispanic men, and that I thought of them as being in general more violent and temperamental than I. Some reading about Hispanic cultural and family life helped me to get a more balanced picture of this group, with whom I had very little contact.

Mary recalls that she was somewhat apprehensive about working with same-sex couples in therapy. She attributed the discomfort to being generally uninformed about the homosexual culture:

> After reading a good deal and attending workshops, it became apparent that my discomfort stemmed from such stereotypical beliefs as gay individuals are more maladjusted, unhappy, and pathological than their heterosexual counterparts. It was only after an exploration of these beliefs that I could better understand the culture, as well as the basis for my own homophobic thinking.

One very important part of you, your feelings, thoughts, and behavior is your culture. The term *culture* is defined in many different ways, but for our purposes, we use the definition offered by Donna Gollnick and Philip Chinn (1990) that culture is a shared and commonly accepted set of beliefs, practices, and behaviors. They go on to point out that there are both microcultures and macrocultures. The macroculture consists of those beliefs and practices shared by the majority of citizens. So, for example, in the United States, a belief in democracy is shared by most (although not all) citizens and would be considered part of the macroculture. Microcultures, on the other hand, are those beliefs and practices that come from membership in a smaller group or subgroup. There are, for example, attitudes and beliefs that are more accepted by women than by men, and vice versa. Therefore, as you get to know yourself, you need to consider your membership in various groups and subgroups in society and the positive and negative associations you have with your own and other subgroups.

You are a member of many subgroups. Some of them are temporary; you can move in and out of them as you choose. For example, you are now a college student, but you could stop being one at any time, and eventually you will no longer be a student. Other subgroups are relatively permanent; your membership in them is determined primarily by accident of birth. Your race, gender, and ethnicity are examples. Your age will change constantly, of course, but you will always be part of an age group. Some have argued that sexual orientation is determined at birth; others believe it can change. However, your status as a heterosexual, homosexual, or bisexual is a relatively permanent feature of your life. Social class is another important subgroup, and here you need to consider both the social class in which you were raised and the one to which you now belong. Your cultural identity consists of these subgroups, the degree of identification you make with them, your status as a member of dominant or subordinate groups, and your stage of subgroup identity development.

Degree of Identification

Knowing the subgroups you belong to is only part of the picture. Another important part is how strongly you identify with that group. There are Jews, for instance, who identify very strongly with their Jewish heritage. They observe the traditions and holidays, obey dietary laws, and travel to Israel. There are also Jews who do none of these things. Individuals make varying degrees of identification with subgroups to which they belong. Although both of us are very aware of issues that confront us as a man or woman, we vary considerably in our identification with other subgroups. Fred, for example, knows almost nothing about his ethnic heritage, which is a mixture of Irish, English, and German, and is aware of very little influence from that subgroup. Mary, on the other hand, identifies strongly with both her Lithuanian and Irish heritage and has a good sense of how those traditions affect her as a person; however, she struggles to appreciate fully the effects of her race (Caucasian) in her personal and professional life.

Dominant vs. Subordinate

It may be that one or more of the groups you listed earlier is often referred to as a *minority group*. This term has been used not only to describe a group's relative numbers, but also its position in society. Thus, women are sometimes referred to as a minority group, although they outnumber men. With regard to racial groups, population projections indicate that by early in the 21st century, members of various racial groups in the United States now referred to as minorities will outnumber Caucasians.

We prefer to use the terms *dominant* and *subordinate* to describe various subgroups (Hardiman & Jackson, 1997). Dominant groups are those who hold the majority of resources and power. They are more often in control of major institutions in society and are more often in positions of power in government, education, and business. So, for example, while women outnumber men in the general U.S. population, the vast majority of CEOs, members of Congress, college presidents, state governors, and so on are men. Thus, with regard to gender, men are the dominant group and women the subordinate. Using the same logic, in the United States, whites are dominant with respect to race, heterosexuals with respect to sexual orientation, upper and upper middle class with respect to social class, and Christians with respect to religion.

If you think about it, you will see that it is a mistake to refer to a person as dominant or subordinate. Most of us are members of some subgroups that happen to be dominant, and others that happen to be subordinate. If you are a white, heterosexual, Jewish female from a working-class background, you are a member of two dominant groups and three subordinate groups. Go back to your list of subgroups and label each one as dominant or subordinate.

If you are hearing these terms in this context for the first time, you may have some trouble accepting them; many people do. After all, if you are a woman, you may not feel subordinate to men, and if you are white, you probably have no desire to dominate people of color. The terms do not necessarily describe your behavior. You are, however, a member of groups that, as groups, are relatively more or less powerful in society. And in general, the experience of members of the dominant group tends to be pretty different from that of subordinate groups.

Dominant groups are afforded certain privileges by society. You may find that hard to believe, given all the publicity that affirmative action and other programs receive, but many of the privileges given to members of the dominant group are subtle. For example, both of us are white, heterosexual, and Christian. We can go into a department store and wander around as long as we want; friends of ours who are Black or Hispanic have told us that they are frequently followed by store detectives or questioned by clerks after only a few minutes. Wherever we have gone to school or to work, the major holiday in our religion is at least 1 day off for us and sometimes 2 weeks! When we are out in public with our partners, we can hold hands, hug, or even kiss without worrying about who may be watching or disapproving.

While members of both the dominant and subordinate groups may have stereotypes or prejudices about members of the other group, because the dominant group has more power its members are more often in a position to enforce those prejudices, sometimes in very subtle ways. For example, one stereotype of women is that they are less assertive than men. For several years, Fred was an administrator at a residential treatment center for emotionally disturbed adolescents. Listen to what he says about his hiring practices:

> Although I would have told you I did not subscribe to that particular stereotype, looking back I see a difference in the way I looked at candidates for supervisory positions. Discipline and assertiveness are very important in those positions, and while I hired both men and women, I realize now that I scrutinized the women much more closely than the men. It was as if the women had to prove to me that they could handle that aspect of the job, while the men had to prove they could not.

Because of these dynamics, members of subordinate groups are subjected to subtle, and not so subtle, acts of prejudice and discrimination. Some members of the dominant group perpetuate these acts deliberately (conscious oppression), while others are not even aware they are doing it (unconscious oppression). The incidents described earlier are more subtle. At the other end of the spectrum are verbal and physical harassment and physical assaults. The dynamics of dominant and subordinate groups are complex, interesting, and important. We encourage you to explore this topic some more. There are suggestions at the end of the chapter on how to do so.

Stages of Identity Development

If you are a member of a subordinate group, you may have gone through a period in your life when you didn't have much of a desire to interact with members of the dominant group. Or perhaps you may never have felt this way, but have been puzzled by others who do. As a member of a dominant group, you may also have been puzzled or hurt by this behavior. You may have thought of yourself as a nonprejudiced person and realized with some shock and shame that several of the things you and members of your group have done were harmful. All these attitudes and feelings are part of different ways that people make sense of their membership in dominant and subordinate groups.

A number of authors have charted stages that people seem to go through as they grapple with these issues. Racial identity development theories appeared first (Hardiman, 1979; Jackson, 1976). Bailey Jackson, Rita Hardiman, and others at the University of Massachusetts have extended these ideas to dominant and subordinate groups in general and written about their application to issues of race, gender, religion, class, sexual orientation, and physical ability (Hardiman & Jackson, 1997). They have identified four stages that members of a subordinate group go through and four that members of a dominant group go through in the development of their subgroup identity. We begin with the stages of the subordinate groups.

SUBORDINATE GROUP IDENTITY STAGES

Stage 1: Acceptance/Conformity People at Stage 1 either actively deny that there is any discrimination or passively accept it as "the way it is." They tend to believe that the dominant group occupies positions of power because its members have skills and abilities that members of the subordinate group do not have. They look to members of the dominant group for approval and validation and react strongly to any signals from a member of the dominant group that they have done something wrong. Approval and support from members of their own group, while appreciated, are less important.

Stage 2: Resistance/Reaction In this stage, members of the subordinate group often reject the norms and values set by the dominant group and try to build a positive sense of themselves as members of the subordinate group.[2] Sometimes this involves an almost total rejection of the dominant group accompanied by feelings of anger, pain,

[2]The term *subordinate* is used to distinguish between the power position of subgroups. In this stage, while building a positive identity, members of the group probably do not refer to themselves as subordinate.

and distrust. A Hispanic, for example, may become more aware of discrimination at the hands of Whites, and dissociate himself with White society as much as possible. He may stop spending time with White friends, co-workers, or with Hispanics who do not see the world the same way, and may confront these people angrily over things they have done or said. He may also take steps to identify and spend time with the Hispanic community where he lives. Having grown tired and resentful of being punished or disapproved for not being White or not speaking, acting, dressing, and eating as a White person does, he may now seek to measure himself against standards drawn only from the Hispanic community and tradition.

Stage 3: Redirection/Redefinition Here the process of building a positive identity continues, but now little energy is put into rejecting the dominant group. Instead, the person concentrates on the movement, begun in the last stage, to identify values and structures that are unique to the subgroup and to define him- or herself as a member of that group. Although the person may withdraw from the dominant group, the angry rejection is gone. The person tends to seek out others who are at a similar stage in their identity development and draw support and validation from them.

Stage 4: Internalization During this stage, the new identity becomes an integral part of the person. Because less energy is required to build and protect this new identity, the person broadens her or his experience and involvement. There is increased interaction with members of the dominant group or with members of the subordinate group who are at different stages in their own identity development. Membership in a particular subordinate group is seen as only one part, albeit an important one, of the person's total identity.

DOMINANT GROUP IDENTITY STAGES

Stage 1: Acceptance/Conformity People at this stage can accept their status either actively or passively. Active acceptance involves conscious acceptance and approval of their superior status. People at this stage also discuss negative stereotypes of the subordinate group actively and appear to believe them to some degree. They deny that they are racist, sexist, or anti-Semitic (and so on) and insist that the subordinate group is to blame for its own problems in the world. Passive acceptance is a bit more subtle. Here people are not aware of their discrimination and prejudice. They may concede that there is some racism or sexism, for example, but that it is blown out of proportion and that they themselves do nothing to support it. They are offended and hurt by charges to the contrary.

Stage 2: Resistance/Reaction Many people in the dominant group remain at Stage 1 for a long time or even throughout their lives. However, others begin to realize that there is prejudice and discrimination in society and in themselves. They see the negative effects that such discrimination has on members of the subordinate group, as well as on members of the dominant group, and often feel shame, guilt, or anger. This can be a very painful period.

Stage 3: Redirection/Redefinition In this stage, the person rebuilds identity as a member of the group, but in a different way. A man may define himself proudly as a man, for example, but not as a sexist. Actively working to eliminate sexism in himself and others, he now looks for the positive aspects of membership in his subgroup. He looks for other men at the same level to support him in his efforts and share his struggle.

Stage 4: Internalization Finally, the new identity is fully integrated and, again, is only one part of the person's total identity. There is increased appreciation for diversity and for people at every stage of identity development.

It will take time to think about your own position on these stage continua. However, stage of identity is an important part of understanding why you react in certain ways to others and why they react to you as they do. We encourage you to read more about specific identity development theories. Recommended resources are at the end of this chapter.

Why have we devoted so much time to the issue of cultural identity? Many of the values, attitudes, and reaction patterns you examined earlier are strongly influenced by your membership in subgroups and the attitudes you learned about your own and other groups. Understanding the sources of these aspects of your personality will help you see them as only one of many possible ways to be in the world.

Furthermore, in your internship and your life, you will surely interact with people whose cultural identity is different from your own. You need to understand your reactions to them and theirs to you. Understanding cultural identity will help you accept others for who they are and avoid displacing feelings or prejudices you have about subgroups onto your clients and co-workers.

SUMMARY

This chapter has encouraged you to begin, continue, or develop the habit of self-examination, and we have suggested specific areas that will be important in your work. Making an investment in self-understanding will help you see your particular issues as you move through the stages of an internship. It will also help you identify and overcome obstacles and deal more effectively with clients, supervisors, and co-workers. Perhaps you will also see the benefits in your life outside the placement. There are some other issues about yourself that you need to explore. These issues, which pertain only to your functioning as an intern, are the subject of the next chapter.

For Further Reflection

1. Review the areas of values discussed in the book. How many of them are areas about which you have strong feelings? Are there other core values for you that are not listed? Have your feelings about any of these issues changed over the years? Why do you think that happened? Discuss your answers with one or two other interns.

2. Which of these values do you think will be important in your internship? How will it feel to encounter others who do not share those values?
3. Think about how you prefer to learn. Do you prefer concrete methods of taking in information or more abstract methods? How about in processing information—do you tend to be concrete or abstract? Give examples. What learning activities and opportunities exist at your internship that are well suited to your style? Are there any that will stretch it?
4. Do you see yourself using primarily separate knowing, connected knowing, or some of both? Give examples of the predominant style(s) in your life. How will your style be a strength and a liability in your internship?
5. Think about recent times when you have found yourself responding to a situation in a way you later regret. Were any of these incidents typical of you? Do you think they are indicative of a response pattern? Are you aware of any other patterns that you struggle with? Try to write them out using the format on p. 23. Which of these patterns might prove a challenge at your internship?
6. What are some of the important rules and roles in your family (both formal and informal)? How do they affect you? In what ways might they affect you in your internship?
7. Choose one of the psychosocial stages described. Think about your experience with the central issue of that stage and with the polarities as well. Where do you think you are on the continuum of outcomes? Has that changed? Where do you see evidence of that position?
8. Make a list of all the subgroups you belong to (race, gender, class, etc.). Next to each one, note whether it is a dominant or subordinate group and rate it 1–5 (low to high), depending on how strongly you identify with that group. Explain each of your ratings.
9. Are there subgroups in society about which you have some stereotypes? Will you be encountering any of these groups in your internship?
10. Can you trace your own journey through the stages of dominant or subordinate identity development?

For Further Exploration

ON THE IMPORTANCE OF SELF-UNDERSTANDING

Brill, N. (1998). *Working with people: The helping process.* (5th ed.). New York: Longman.

Excellent chapter on self-understanding.

Corey, M. S., & Corey, G. (1998). *Becoming a helper* (3rd ed.). Pacific Grove, CA: Brooks/Cole.

A very readable and thought-provoking book, devoted entirely to self-understanding and effectiveness in interpersonal relationships.

ON VALUES:

Corey, G., Corey, M. S., & Callanan, P. (1998). *Issues and ethics in the helping professions.* (5th ed.). Pacific Grove, CA: Brooks/Cole.

The issue of values is woven throughout this book, but one chapter in particular discusses the issue of imposing-vs.-exposing values.

ON REACTION PATTERNS

Weinstein, G. (1981). Self science education. In J. Fried (Ed.), *New directions for student services: Education for student development* (pp. 73–78). San Francisco: Jossey-Bass.

Describes an approach to uncovering and interrupting these patterns.

ON LEARNING STYLE:

Claxton, C. S., & Murrell, P. H. (1987). *Learning styles: Implications for improving educational practices* (ASHE-ERIC Higher Education Report No. 4). Washington, DC: Association for the Study of Higher Education.

An overview of many different kinds of learning style theories.

Kolb, D. A. (1984). *Experiential learning: Experience as the source of learning and development.* Upper Saddle River, NJ: Prentice Hall.

Kolb's basic ideas explained.

Kolb, D. A. (1985). *Learning style inventory.* Boston: McBer.

A way to see where your learning preferences are.

ON SEPARATE VS. CONNECTED KNOWING

Belenky, M. F., Clinchy, M., Goldberger, N. R., & Tarule, J. M. (1986). *Women's ways of knowing: The development of self, voice and mind.* New York: Basic Books.

An excellent and thorough treatment of this topic, with summaries of other research as well.

Gilligan, C. (1982). *In a different voice.* Cambridge, MA: Harvard University Press.

This groundbreaking work laid the foundation for much of the work on women's development.

ON FAMILY PATTERNS

Goldenberg, I., & Goldenberg, G. (1996). *Family therapy: An overview* (4th ed.). Pacific Grove, CA: Brooks/Cole.

Extremely readable text on families and family therapy.

Goldenberg, I., & Goldenberg, H. (1996). *My self in a family context: A personal journal* (4th ed.). Pacific Grove, CA: Brooks/Cole.

The accompanying workbook serves as a guide for you to explore your own family dynamics.

ON USING ERIKSON

Corey, G., & Corey, M. (1997). *I never knew I had a choice* (5th ed.). Pacific Grove, CA: Brooks/Cole.

An excellent book on self-understanding based on Erikson and other theorists.

Sweitzer, H. F. (1993). Using psychosocial and cognitive behavioral theories to promote self-understanding: A beginning framework. *Journal of Counseling and Human Service Professions*, 7(1), 8–18.

Combines Erikson and reaction patterns and discusses implications for human service work.

ON CULTURAL IDENTITY

Adams, M., Bell, L. A., & Griffin, P. (Eds.). (1997). *Teaching for diversity and social justice.* New York: Routledge.

A collection of readings that describes workshops on many forms of oppression. Useful as a resource for trainers with exercises to work through.

Green, J. W. (1995). *Cultural awareness in the human services.* Upper Saddle River, NJ: Prentice Hall.

Covers both theoretical and practical aspects of working with diverse populations.

Ivey, A. E., Ivey, M. B. & Simek-Morgan, L. (1993). *Counseling and psychotherapy: A multicultural perspective* (3rd ed.). Needham, MA: Allyn & Bacon.

A relatively new version of a text that has been around for a long time. Unlike some other texts, where multicultural issues have just been tacked onto each chapter, Ivey et al. have fully integrated this perspective into a basic counseling text.

References

Adams, M., Bell, L. A., & Griffin, P. (Eds.). (1997). *Teaching for diversity and social justice.* New York: Routledge.

Axelson, J. A. (1990). *Counseling and development in a multicultural society* (3rd ed.). Pacific Grove, CA: Brooks/Cole.

Belenky, M. F., Clinchy, M., Goldberger, N. R., & Tarule, J. M. (1986). *Women's ways of knowing: The development of self, voice and mind.* New York: Basic Books.

Brammer, L. M. (1985). *The helping relationship: Process and skills* (3rd ed.). Upper Saddle River, NJ: Prentice Hall.

Brill, N. (1998). *Working with people: The helping process.* (6th ed.). New York: Longman.

Claxton, C. S., & Murrell, P. H. (1987). *Learning styles: Implications for improving educational practices* (ASHE-ERIC Higher Education Report No. 4). Washington, DC: Association for the Study of Higher Education.

Corey, G., & Corey, M. (1997). *I never knew I had a choice* (6th ed.). Pacific Grove, CA: Brooks/Cole.

Corey, G., Corey, M. S., & Callanan, P. (1998). *Issues and ethics in the helping professions* (5th ed.). Pacific Grove, CA: Brooks/Cole.

Corey, M. S., & Corey, G. (1998). *Becoming a helper* (3rd ed.). Pacific Grove, CA: Brooks/Cole.

Erikson, E. H. (1963). *Childhood and society.* New York: Norton.

Gilligan, C. (1982). *In a different voice.* Cambridge, MA: Harvard University Press.

Gilligan, C., Ward, J. V., & Taylor, J. (1988). *Mapping the moral domain.* Cambridge, MA: Harvard University Press.

Gollnick, D. M., & Chinn, P. C. (1990). *Multicultural education in a pluralistic society* (3rd ed.). Columbus, OH: Merrill.

Hardiman, R. (1979). *White identity development theory.* Unpublished doctoral dissertation, University of Massachusetts.

Hardiman, R., & Jackson, B. W. (1997). Conceptual foundations for social justice courses. In M. Adams, L. A. Bell, & P. Griffin (Eds.), *Teaching for diversity and social justice* (pp. 16–29). New York: Routledge.

Jackson, B. W. (1976). *Black identity development theory.* Unpublished doctoral dissertation, University of Massachusetts.

Kegan, R. (1982). *The evolving self: Problem and process in human development.* Cambridge, MA: Harvard University Press.

Kegan, R. (1994). *In over our heads: The mental demands of modern life.* Cambridge, MA: Harvard University Press.

Kolb, D. A. (1984). *Experiential learning: Experience as the source of learning and development.* Upper Saddle River, NJ: Prentice Hall.

Kolb, D. A. (1985). *Learning style inventory.* Boston: McBer.

Kolb, D. A., & Fry, R. (1975). Toward an applied theory of experiential learning. In C. Cooper (Ed.), *Theories of group process.* New York: Wiley.

Krueger, D. (1984). *Success and fear of success in women.* New York: Free Press.

Lyons, N. (1983). Two perspectives on self, relationships and morality. *Harvard Educational Review,* 53, 125–145.

Schenkel, S. (1984). *Giving away success.* New York: McGraw-Hill.

Schram, B., & Mandell, B. R. (1997). *An introduction to human services* (3rd ed.). New York: Macmillan.

Stewart, G. M. (1990). Learning styles as a filter for developing service learning interventions. *New Directions for Student Services, 50,* 31–42.

Sugarman, L. (1985). Kolb's model of experiential learning: Touchstone for trainers, students, counselors and clients. *Journal of Counseling and Development, 64*(5), 264–268.

Sweitzer, H. F. (1993). Using psychosocial and cognitive behavioral theories to promote self-understanding: A beginning framework. *Journal of Counseling and Human Service Professions,* 7(1), 8–18.

Sweitzer, H. F., & Jones, J. S. (1990). Self-understanding in human service education: Goals and methods. *Human Service Education, 10*(1), 39–52.

Weinstein, G. (1981). Self science education. In J. Fried (Ed.), *New directions for student services: Education for student development* (pp. 73–78). San Francisco: Jossey-Bass.

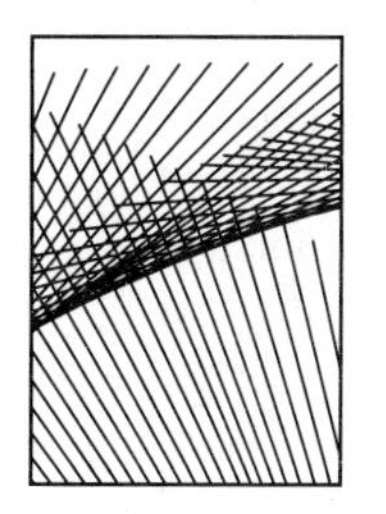

CHAPTER 3

Understanding Yourself as an Intern

If I can't help someone with a problem, I feel like a failure. I guess I feel like I should understand everyone and I should know what to do in every situation.

STUDENT JOURNAL ENTRY

In the last chapter, we encouraged you to examine who you are in general terms and discussed how some of what you discovered might be relevant to your internship. No doubt you discovered or rediscovered some issues that are relevant to your life and your placement and some that are not. In this chapter, we ask you to consider some specific issues that will be important as you begin your internship. Your motivation for your work, along with any unresolved or partially resolved issues that you bring to it, are important issues to consider. We will also ask you to consider your feelings about self-disclosure, assessment, and authority, which we have found to be features of nearly every internship. Another issue that many interns don't think enough about is life outside the placement, which will surely affect, and be affected by, the work that you do. Finally, we ask you to examine and assess your personal support system, which will be an invaluable asset to you throughout your placement.

YOUR MOTIVATION FOR PLACEMENT

At some point in your education, someone has probably asked you to think or write about the reasons you are considering human services as a career. Perhaps, too, you have been asked by a placement coordinator or a prospective placement why you are interested in a particular kind of internship. Understanding and reminding yourself of these motivating factors can be a source of strength. Yet every one of them can be a liability as well. Corey and Corey (1998) discuss several possible reasons for entering human services, or needs that workers bring with them, including the need to have an impact, the need to care for others, the need to help others avoid or overcome problems they themselves have struggled with, the need to provide answers, and the need to be needed. They also point out that each one of these motivations can cause problems.

There is an important distinction here between wanting and needing. All the reasons just listed are good ones for entering the human services field, provided you substitute "want" for "need." For example, wanting to care for others is fine, but as Corey and Corey (1998) point out, it is not fine if you always place caring for others above caring for yourself. Sometimes what you need conflicts with what the placement wants or needs. For example, you have worked a very full week and are very tired. Your supervisor explains that someone is out sick and asks if you would mind working over the weekend. If you choose to say "yes" to these requests from time to time, that is fine. However, if you find that you cannot say "no," ever, then you may be someone who needs to put others' concerns first in order to feel good about yourself. That can happen for many reasons, but none of them offset the price you will pay if you don't learn to attend to your own needs as well as those of your clients.

Providing answers is great, too; coming up with a solution for a problem is a wonderful feeling. However, there are going to be problems you cannot solve. Furthermore, there will be times when it is better to let someone struggle to find an answer than to jump in with a lot of advice, even if your advice is on target and would solve the problem faster. If you need to be the one with the answers, these situations are likely to be difficult for you. Take some time to think about your own motivations for entering the sort of placement you have chosen. If those desires turn into needs, what kinds of problems could that create for you?

UNRESOLVED ISSUES

Each of us has struggled with different personal issues in our lives, and the work you did on your psychosocial identity in the last chapter may have uncovered or reminded you of some. There are other kinds of issues that can linger as well. You may have had a prolonged struggle with one of your parents during adolescence. You may have been a victim of abuse or assault. Perhaps you wrestled with substance abuse or an eating disorder. These struggles can leave us all with issues that are unresolved or partially resolved.

For example, you may have overcome your eating disorder, but the memory of those days may still be very painful. These issues could be thought of as your "unfinished business." This unfinished business does not have to come from some traumatic event or a particular struggle such as those just mentioned. Family patterns are another source. For example, if you were always the "mediator" in your family—the one who stepped in and calmed people down and helped them resolve their differences—you may have some very strong feelings about conflicts. It may be hard for you to see someone in a conflict and not step in to help, even though it is sometimes best to let people work it out for themselves.

As we touched on in the last chapter, your unfinished business can also come from your membership in certain societal subgroups. Your vulnerable areas are not just the result of your personality or of your childhood and family experiences. They are also the result of your experience in society as a member of racial, ethnic, gender, and other subgroups. If you are a member of a group that has been discriminated against, you may have a difficult time with clients who express prejudice, especially toward that group. If you are a member of a dominant group, such as males, Whites, or heterosexuals, and have thought about issues of discrimination, you may feel guilty, or very hesitant, around members of oppressed groups.

The point here is not that you should be free of unfinished business; all of us have some. However, you need to be as aware as you can of what that business is and how those areas of vulnerability may be touched in your work. Often, past struggles are the reason people choose human service work (Collins, Fischer, & Cimmino, 1994). They want to help others deal with or avoid problems that they experienced and feel that their experience will be an asset. In many cases, it is. However, as we alluded to in the last chapter, some interns try to use the placement to resolve issues that they themselves have not yet resolved. If you encourage, or insist, that a client do what you cannot (e.g., be assertive or show love) or rush to protect a client from a critical parent (in a way that you never were), you are falling into this trap.

Even if your unfinished business is not part of the reason you selected human services as a career, human service work can often stimulate those issues. Suppose you are working with a substance abuse population. As you work with a particular client, he begins to discuss his struggle with an overcritical father. Even though you have not had experience with being a substance abuser, you have had a similar struggle with your father or some other parent figure. You are likely to be touched by this client in ways that seem mysterious at first. You may find yourself thinking about him when you are at home, or when you wake up in the morning. This kind of preoccupation can be very draining.

In addition, some clients have an uncanny sense for your sore spots and will use them to manipulate you. If you have a strong need to be liked, for example, the client may withhold approval as a way to get you to relax a rule or go along with a rationalization. Your unfinished business can show up with colleagues and supervisors as well. If your father or mother was hypercritical of you, you may decide that in the internship you are not going to accept criticism without standing up for yourself.

SELF-DISCLOSURE

All of us set boundaries around certain areas in our lives. There are aspects of ourselves we would discuss with anyone and aspects we would discuss only after we know someone. An even more private "zone" consists of those things we would only tell our closest friends. There may even be things about yourself, such as events in your past or feelings you struggle with, that you would not discuss with anyone. How did these zones come to be private? Surely you have noticed that there are people who seem hesitant to discuss something that is perfectly comfortable for you or who are very open about aspects of themselves that you consider private. Perhaps it is your family that taught you where to draw the line. Perhaps it is your culture or subculture. It may also be that you had experiences earlier in your life that left you cautious about trusting people with information about yourself. Regardless of where the boundaries come from, it is important to know where they are and how rigid they feel. At the end of this chapter is an exercise, adapted from the work of Paul Pedersen (1988), that can help you clarify these issues.

The work at your internship is likely to challenge your boundaries. Clients may try to find out personal information about you for a variety of reasons. Some may just want to get to know you better so they can feel comfortable with you. In some cultures, it is very important to exchange some personal information before addressing a problem or asking for help. Furthermore, going to someone for help, or especially being a client in a residential situation, can make a person feel powerless. Getting some information about you is a way to tip the balance of power a little more in the client's direction. Finally, some clients are actively looking to put you off balance.

You may feel hesitant to discuss a particular area but wonder whether you should try to overcome that reluctance. Or you may feel very comfortable but wonder whether certain topics are appropriate in a particular setting. For example, one study found that clients had little interest in the attitudes of their counselors but were interested in knowing about personal feelings, interpersonal relationships, professional issues, and successes and failures (Hendrick, 1990). In any case, you need to be prepared and think about what you will and will not discuss. Self-disclosure can be an issue with coworkers, as you decide how much of yourself to share and with whom. Finally, self-disclosure is important in supervision. You will need to think about, and possibly negotiate, how much of your thoughts and feelings you want to share and what is expected of you.

ASSESSMENT

You have been assessed and given feedback in some form in many areas during your life, including academic, artistic, and athletic. It probably began sometime during school. Most people get at least a little nervous when someone else judges their efforts, but for some people, it creates a great deal of anxiety. You may have been successful in some, or all, assessment areas in the past. You may also have a strong sense of competence, which you will recall is only partially related to success. If so, being assessed is probably relatively comfortable for you. However, it may be that you struggled in one or more of those areas or that you don't have a particularly strong sense of competence.

Some of you have worked very hard to achieve the level of academic success you now enjoy. You have learned how to be successful in school and how to adapt to the different demands and styles of your professors. Even so, you are entering a new area now. The attitudes, skills, and knowledge required for success at your placement site may be quite different from those required to be successful in classes. There is also likely to be an emphasis on what you can do, rather than what you know.

If you have had other field experiences prior to your internship, the transition will be less dramatic, although each placement is different. For some interns, the internship feels like an area in which they have to prove themselves all over again (Kaslow & Rice, 1985; Lamb, Baker, Jennings, & Yarris, 1982). Think about your feelings and experiences with being assessed. Does the prospect of a new round of assessment make you nervous? Do you know why? Is there anything you can do to ease your anxiety?

AUTHORITY

If you are working with clients, you may be in a position of some authority. For those of you who are parents, older siblings, former camp counselors, and so on, this is old news. For others, though, it will be a new experience, and it is one that causes many interns some concern (Blake & Peterman, 1985). Exercising authority isn't easy, and sometimes the people on the other end don't appreciate it. Being an authority figure can be especially difficult when your clients are close to you in age or have similar life experiences or circumstances. For example, we have had many interns of traditional college age who worked with adolescents just a few years younger. Their own adolescence was still fresh in their minds and with it the desire to test limits. Some had to struggle to feel comfortable enforcing agency policy with their clients. Think about your own experiences with authority, both as the authority figure and as the recipient of authority. How easy do you think this is going to be? Look back over what you have been learning about your life experience. Are there certain populations or issues that you may find difficult?

YOUR LIFE CONTEXT

> *My biggest concern is stress, basically from outside the internship. The stress to get all the papers done on time, working many hours a week at my job to pay bills, and still put in 30 hours in the internship.*
>
> STUDENT JOURNAL ENTRY

Your life context consists of all the other things going on in your life besides the internship. That context will vary according to your family situation, your social life, and the configuration of your academic program. You will have some responsibilities outside your placement, some expectations placed on you (or being continued), and possibly some stress. Here are some areas to consider.

School

In some programs, interns take very few classes during their internship semester. But in others, the internship can be the equivalent of only one course, leaving a full-time student with three or four other courses to carry. Some students even need to carry an overload because they must finish school in a certain time frame. The internship tends to be at least as demanding of your time as one course, or even two. Our students also report that internships demand a good deal of psychological and emotional energy. The work can be exhilarating, and many interns find themselves thinking about their work when they are not at the placement, or putting in extra time to read and research issues relevant to their work. But the work can also be emotionally draining, and the clients' situations can be heartbreaking. Many interns have told us that they find themselves thinking and talking about the internship a great deal.

Work

Some internships are paid; most are not. If you have been holding down a job in addition to going to school, you will want to think about whether you can and should give it up during the placement. Some interns cannot do this; their economic situation simply prohibits it. If you are going to be working, think about the schedule you will have and how it will fit into the time demands of your placement and coursework.

Roommates

If you live with other people, they are part of your life context. Schedules for housekeeping, sleep, quiet time, study, and entertainment are areas of negotiation with roommates. However, you may need to renegotiate these issues depending on the demands of your placement.

Family

In this category we are including your family of origin (your parents, siblings, and anyone else who lives with them) and your nuclear family (your spouse or partner and children). You may be living with any, all, or none of these people, but you undoubtedly have responsibilities to them. What are they? How flexible are they? Could they be changed for this semester if the demands of the placement warrant it?

Intimate Partners

If you have a special partner in your life, even though you may not live with that person, this is another area of responsibility. It took time and energy to build that relationship, and it will take at least some time and energy to maintain and nurture it. How much time have you been devoting to that relationship? What does that person expect of you, especially now that you have begun your placement?

Friends

Friendships are responsibilities, too. Some friends are closer than others. You may spend time with them individually or in groups. How much of your time do you spend with friends? How much time do you hope to spend with them this semester? What demands do they put on your time and energy?

Yourself

How much time and energy do you currently devote to taking care of yourself? By taking care of yourself, we mean sleep, good meals, exercise, and hobbies. All of us have activities we engage in for pleasure and as a source of stress reduction. Hiking, playing the piano, painting, lifting weights, meditating, and reading are just a few examples.

You will need to take the best care of yourself that you can to meet the demands of the internship. You may be able to cut down some on hobbies if necessary, but you will need to be realistic about the time required to take proper physical and emotional care of yourself.

Doing It All

Many interns make the mistake of just inserting an internship into an already busy life and expecting it all to work out somehow. Some of our busiest students fall prey to this misconception, especially those raising families, working full time, or seriously involved with sports. We are going to encourage you to look a little more objectively at your life context. If you list all the people and activities that you will be committed to this semester and estimate the number of hours each week or month each of those commitments is going to take, you may be shocked to discover that you have committed yourself to more hours than exist in a week or a month. If so, just as when you are doing your financial budget, you will have to make some compromises. Often, this compromising involves talking with others to whom you now have commitments about reordering things, just for the short run.

Many interns report feeling that they have less time than they ever had. When we ask them to inventory their commitments, as just described, they find that they should have enough hours in the day and week, but they just do not feel as if they do. Sometimes that comes from poor estimating. For example, if a class lasts for an hour, you need to remember that you will spend some time getting to and from class. However, another reason for this feeling is that interns often underestimate the psychological commitment and emotional demands of the internship. Whether you realize it or not, your emotional energy is taxed by an internship, and that is going to affect how available you feel for other activities and responsibilities.

SUPPORT SYSTEMS

Balancing all the demands we were just discussing is your support system, which is made up of people who give you what you need to get through life's challenges. Charles Seashore (1982), who has written extensively about support systems, calls them "a resource pool drawn on selectively in order to support me in moving in a direction of my choice that leaves me stronger" (p. 49). Your support system will be an important part of helping you meet the demands of your internship and the other demands in your life. As one of our interns pointed out in a journal entry, "Support teams are a way to relieve stress and frustration . . . they listen to you, give advice, and give support. It is very comforting to know that you have someone to fall on when needed."

Everyone needs some support. Some people need more than others, and you will need more at some times than other times, but you do need it and you will continue to need it. Sometimes, people in the helping professions have trouble accepting that they need support. Hence, they do not seek it out or decline when it is offered. They give to others unstintingly and enjoy being people that others can count on, but they are better at giving than receiving. Most of the time, sooner or later, they become exhausted and are of little use to anyone. It may be that they have an image of the perfect helper—someone who never needs help. "How can I help others," students have asked us, "if I have problems myself?" Others know they need support, but they seem unable to accept it. For them, this may be a dysfunctional pattern of the sort described in Chapter 2. Accepting your need for support and developing a strong support system are important parts of becoming a human service professional.

Your support system is made up of many different people, and you will need different kinds of support at different times (Seashore, 1982). Here is a partial list of the kinds of support you might need. You may be able to add to it.

Listening Sometimes you just want someone who will listen to you without criticizing or offering advice. The person listening should be someone to whom you can say almost anything and on whom you can count not to grow restless or frightened. Think of these people as your "sounding boards."

Advice On the other hand, sometimes you need sound advice. You may not always follow it, but you need a source of advice that you can trust. Think of these people as your "personal consultants."

Praise There are times when what you need most is for someone to tell you how great you are. If they can be specific, all the better. Think of these people as your "fans."

Diversion Some people are friends you can count on to go out and play with. You don't have to talk about work, your problems, or anything else. You just have activities you enjoy together. Think of these people as your "playmates."

Comfort When we were children and we became ill, there was nothing we wanted more than pure comfort. A comfortable place to rest, good food, music we enjoy, all without having to lift a finger! At times, ill or not, this is still just the kind of support we need. Think of the people in your life who comfort you as your "chicken soup people."

Challenge There are times when challenge is the last thing you want. At other times, though, someone who will push you to do more, look at things in a different way, and confront problems or inconsistencies in your thinking is the best friend you have. Think of these people as your "personal coaches."

Companionship It is good to have people in your life with whom you feel so comfortable that you can do anything, or nothing, with them. Sometimes you may not care what you are doing, only that you are not doing it alone. Think of these people as your "buddies."

Affirmation Another kind of support comes from people who have some of the same struggles that you do. It is a lonely and depressing feeling to believe, or suspect, that you are the only one who is troubled by a particular issue or set of circumstances. Knowing that others feel the way you do, even if they can't change it, can be very helpful. Sometimes there is no substitute for someone who has been through what you are going through. Think of these people as your "comrades."

As we said earlier, you are not going to need all these things at the same time. You are also not going to be able to get all of them from the same person or group. As you read the list, you must be able to think of people who are helpful with one thing but not with others. Both of us have friends who are fun to be with and provide wonderful diversion, but they do not listen well at all! You will know soon enough if you have called on the wrong person for support. For example, you may need good advice, but call on the listener. Even if there were one person in your life who could provide all these kinds of support, that person would soon become exhausted if you asked him or her to meet all your support needs. Similarly, some people who can provide a particular kind of support may not be available at any given time. Try not to become frustrated. You need to learn about your needs and about how different people can best help you meet them.

Think about these categories and add some of your own. Now think about how well each of those needs is currently met in your life. You may find that there are some gaps that you need to address. Your support system will be stronger if you have more than one person in each category. Remember that the internship may, by itself or in combination with other things occurring in your life, create a greater need for support than you now have. The internship may also strain your existing support system. If most of your friends are not involved in internships, for example, they are not going to know how you feel, and they may not understand what you do or how difficult it is. "That's work?" they may say, or "You get credit for that?" They may not understand your need to get to sleep early, work on weekends, or cut back on your social life.

If you have discovered gaps in your support system, we urge you to take steps to fill them. Cultivate new friends; discuss your needs with friends and family. Like investing in self-understanding, it will pay off for you now, and later.

SUMMARY

These last two chapters have asked quite a bit of you. You have looked at yourself as a person and as an emerging professional through a number of lenses. Your assessment of your motivation, your unfinished business, and your response to several common internship issues took time and effort, but they are just as important to carry with you into the internship as any set of helping skills you have learned in class. We also hope you are now in the habit of considering yourself in the context of the people and systems that make up your life. Many human service students learn to assess clients in that way; we have encouraged you to shine that light on yourself as well. Now that you are thoroughly grounded in your individuality, we turn to our theory of stages, which will help you in general terms know what to expect.

For Futher Reflection

1. What are the major reasons you chose human services as a field? This internship in particular? Can you see any way that those motivations could be troublesome for you in your internship?
2. What unfinished business do you think you have at this time? How might that business arise during your internship?

3. Here is an exercise about self-disclosure adapted from Paul Pedersen (1988). For each category listed, mark it as either an area you would talk easily about with relative strangers *(Public)* or one you would only discuss with those close to you *(Private)*:

	PUBLIC	PRIVATE
1. My personal religious views	______	______
2. My views on racial integration	______	______
3. My views on sex and morality	______	______
4. My tastes in food	______	______
5. My likes and dislikes in music	______	______
6. My favorite reading matter	______	______
7. The kinds of movies and TV shows I like best	______	______
8. The kinds of parties and gatherings I enjoy	______	______
9. The shortcomings I have that I feel prevent me from achieving what I want	______	______
10. My professional goals and ambitions	______	______
11. How I really feel about the people I work and go to school with	______	______
12. How much money I make	______	______
13. My general financial situation	______	______
14. My family's finances	______	______
15. Aspects of my personality I dislike	______	______
16. Feelings I have trouble expressing	______	______
17. Feelings I have trouble controlling	______	______
18. Facts about my present sex life	______	______
19. My relationship with my intimate partner	______	______
20. My relationship with my family	______	______
21. Things I feel ashamed of	______	______
22. Things about me that make me proud	______	______
23. How I wish I looked	______	______
24. My feelings about parts of my body	______	______
25. My past illnesses and treatment	______	______

We encourage you to share your answers with others. If you want to keep the answers anonymous, put a number or symbol on the paper that identifies it as yours and exchange papers with classmates. Discuss the similarities and differences you find. Where do you think those differences come from? How do you think your clients might answer these questions?

4. In what ways and in what areas have you been assessed in your life? In how many of those areas were you successful? How do you feel about entering an arena in which you are relatively untested?

5. Review the section of this chapter titled Doing It All. Make a list of all the people, activities, and other responsibilities in your life now (don't forget to include activities to take care of yourself, such as eating, sleeping, and exercise). Now make an estimate of how much time each week you need to devote to each one. Finally, add up the time. If it exceeds 168 hours, you are over budget—that's all the hours there are in a week. If you are over the limit, what do you plan to do about it?
6. Using the following chart, list the people you can count on in each category of support. There are spaces at the end to add your own categories. Now review your chart and see whether there are gaps—spaces with no one—or people who are in danger of being overburdened.

Category	Source 1	Source 2	Source 3	Source 4
Listening				
Advice				
Praise				
Diversion				
Comfort				
Challenge				
Companionship				
Affirmation				
Other ________				

For Further Exploration

Corey, M. S., & Corey, G. (1998). *Becoming a helper* (4th ed.). Pacific Grove, CA: Brooks/Cole.

Good coverage on motivation for going into the helping professions and on unfinished business.

Seashore, C. (1982). Developing and using a personal support system. In L. Porter & B. Mohr (Eds.), *Reading book for human relations training* (pp. 49–51). Arlington, VA: National Training Laboratories.

Theory and application of support systems. A classic article.

References

Blake, B., & Peterman, P. J. (1985). *Social work field instruction: The undergraduate experience.* New York: University Press of America.

Collins, R., Fischer, J., & Cimmino, P. (1994). Human service student patterns: A study of the influence of selected psychodynamic factors on career choice. *Human Service Education, 14*(1), 15–24.

Hendrick, S. S. (1990). A client perspective on counselor disclosure (brief report). *Journal of Counseling and Development, 69,* 184–185.

Kaslow, N. J., & Rice, D. G. (1985). Developmental stresses of psychology internship training: What training staff can do to help. *Professional Psychology Research and Practice, 16*(2), 253–261.

Lamb, D. H., Baker, J. M., Jennings, M. L., & Yarris, E. (1982). Passages of an internship in professional psychology. *Professional Psychology, 13*(5), 661–669.

Pedersen, P. (1988). A *handbook for developing multicultural awareness.* Alexandria, VA: American Association for Counseling and Development.

Seashore, C. (1982). Developing and using a personal support system. In L. Porter & B. Mohr (Eds.), *Reading book for human relations training* (pp. 49–51). Arlington, VA: National Training Laboratories.

CHAPTER 4

The Stages of an Internship

Internship is like a diamond, in that it is multifaceted; it is also like a roller coaster with its highs and lows.

STUDENT JOURNAL ENTRY

Each intern's experience is unique, and yours will be, too. You may have a different experience from other interns at the same placement or from any previous field experiences you have had. If reading the last chapters has accomplished anything, it has helped you see the many aspects of yourself that contribute to the internship experience. It may also have helped you see all the ways that interns differ from one another. Placement sites differ, too, and you may be in a seminar with people who are doing very different work, with very different groups of people.

We have been amazed and enriched by the diversity of experiences that interns have; it is one of the factors that makes working with interns gratifying, even after many years. However, over time, we have also noticed some similarities that cut across various experiences. Some of the concerns and challenges that interns face seem to occur in a predictable order. Our experience, plus our study of other stage theories, have yielded our own theory of internship stages (Lacoursiere, 1980; Sweitzer & King, 1994, 1995). Each stage has its own obstacles and its own opportunities. There are concerns you will have at each stage, and to some extent, those concerns must be resolved for you to move forward and continue learning and growing. However, the process of resolving the concerns is also a learning experience in and of itself. We cannot predict how quickly you will move through the stages; we can only predict the order. At each stage, there are important tasks that will help you address the concerns. If these tasks go undone and the concerns are left unresolved, you can become stuck in a stage.

There are five stages of an internship: anticipation, disillusionment, confrontation, competence, and culmination. The stages are not completely separate; concerns from earlier and subsequent stages can often be seen. However, certain concerns and issues are liable to be particularly prominent during each stage.

An important distinction in understanding the stages is that between morale and task accomplishment (Lacoursiere, 1980). The term *morale* refers to the interpersonal and intrapersonal tone of your experience at the agency. High morale is characterized by positive feelings about yourself, your work, and the agency. The tone is one of hope, optimism, and enthusiasm, and there is movement toward goals, even in the face of obstacles. As much as you would probably like to have high morale at all times, that is not usually what happens. The good news is that morale can often be recovered when it drops, and there is great learning in the process that only occurs if you fully experience both the drop and the recovery.

The term *task accomplishment* refers not so much to the specific tasks assigned by the placement site, but rather to the attitudes, skills, and knowledge that you hope to acquire. Of course, there may be considerable overlap between the two. Here again, you might hope that the growth of this dimension would follow a steady, linear, upward path, but both our experience and some research suggest that this is not the case (Blake & Peterman, 1985). There will be periods when you are learning and growing at an incredible pace. There will also be periods where you feel stuck, and you may be tempted to think you aren't ever going to get where you want to go or that you aren't learning anything. You are always learning, though, or at least the opportunity is always there, and paying attention to what you are learning, rather than dwelling on what you are not, will help you get back on track.

As we mentioned earlier, your rate of progress through the stages is affected by many factors, including the number of hours spent at the agency; previous internships or field experiences; your personality; the personal issues and levels of support you bring into the experience; the style of supervision; and the nature of the work, including the emotional issues stimulated by a particular client population.

STAGE 1: ANTICIPATION

As you look forward to and begin your internship, there is usually a lot to be excited about. Interns often look forward to the internship for several semesters, and it is your best chance to actually get out there, do what you have wanted to do, and make a contribution to others. For most interns, however, along with the eagerness and hope there is inevitably some anxiety. It may not be very visible, even to you, but there are enough unknowns in the experience to cause some concern and anxiety in anyone.

For interns, this anxiety generates the first set of concerns, which generally center on the self, the supervisor, clients, and co-workers. We often refer to this as the "What if . . . ?" stage because interns wonder about things like: What if I can't handle it? What if they won't listen to me? What if they don't like me? or What if my supervisor thinks I know more than I really do? You will probably be concerned about what you will get from the experience and what it is really like to work at this particular site. Many interns wonder whether they can "really do this" and what will be expected of them. Some interns report fears that they are not competent, that they have gotten this far only by great luck, and that in their internship they will surely be found out. You may also wonder about your role; you are not in a student role while at the placement, but you are not a full-fledged staff member either. Depending on your personal situation, you may also be concerned about your family and the effect that such a demanding experience will have on them.

You are going to interact with a number of people during your internship, and it is natural to wonder what to expect from them and whether they will accept you in your new role. You may, for example, be unsure of the role and responsibilities of the site supervisor, and you will be unusual if you don't wonder what your supervisor will think of you and whether he or she will care about you. Those working directly with clients inevitably wonder about how they will be perceived and accepted by clients and just what kinds of behaviors and problems clients are going to exhibit. Finally, most interns are concerned about the reception and treatment they will receive from agency staff members. You may also wonder how you are going to manage the other responsibilities in your life and who is going to be there to support you.

The level of task accomplishment at this time is often relatively low, meaning that you may not be learning the specific things you went there to learn, and that can be frustrating. What is most important at this stage, however, is that you learn to define your goals clearly and specifically and begin considering what skills you will need to reach them. You must also develop a realistic set of expectations for the experience. Since you have not yet actually experienced the internship, but have probably thought and maybe heard a lot about the agency where you will be working, it is inevitable that you will make assumptions, correctly or incorrectly, about many aspects of the internship (Nesbitt, 1993). Some of these assumptions come from stereotypical portrayals in the media of certain client groups (such as the mentally ill) or agencies (such as detention centers); others may come from your own experience with certain issues or problems. As much as possible, these assumptions and expectations need to be made explicit and then examined and critiqued. Finally, you need to work on being accepted by and developing good relationships with supervisor, co-workers, and clients.

STAGE 2: DISILLUSIONMENT

Sooner or later, you are probably going to reach a time when you are not as certain or as positive about your internship as you would like to be. You may find that you are having some trouble getting up and going to the internship or that you are mumbling under your breath or complaining to friends. It is an unusual intern who does not experience some kind of disappointment at some point. One reason for this change is that there is almost always a difference between what you anticipated about your internship and what you really experience.[1] The size of this gap, and hence the dip in your morale, will depend on how successfully you accomplished the tasks of the anticipation stage, but it cannot be avoided altogether. If the concerns of the anticipation stage have been addressed, you will be less likely to encounter a wildly different

---

[1]In some cases, the intern's primary focus is not the work of the internship but the location. For example, some students use an internship in part as a way to be in a major city, like New York or Washington, DC, or even to go abroad. In these cases, disillusionment, if it occurs, is likely to focus on some of the harsh realities of life in a new place. Of course, some interns are focused equally on the work and the location. For them, there may be two possible sources of disillusionment.

reality from what you expected, but there will be some discrepancies and some of them will be troubling. Furthermore, issues will arise that you simply never considered. For some of you, the change will be subtle and barely noticeable; for others, it will be profound and overwhelming.

If anticipation was the "What if . . . ?" stage, disillusionment is the "What's wrong?" stage. Concerns at this stage center on many of the same areas as earlier in the placement: clients, supervisor, the agency, the "system," or yourself. Feelings associated with these concerns may include frustration, anger, sadness, disappointment, and discouragement, and you may find yourself directing any or all of those feelings at the site supervisor, your instructor, clients, co-workers, or even yourself.

This stage is the onset of what we refer to as a crisis of growth. It is possible to become stuck in this stage, and that can have unfortunate consequences. At best, learning and growth will be limited; at worst, the placement may have to be renegotiated or even terminated. On the other hand, letting yourself feel the impact of these issues and working through them present tremendous opportunities for personal and professional growth.

STAGE 3: CONFRONTATION

As the saying goes, "The only way around is through." The way to get past the disillusionment stage is to face and study what is happening to you. Some interns resist acknowledging any problems, even when their level of task accomplishment is dropping. You may fear that any problems must somehow be your fault or that you will be blamed for them. You may think that "really good" interns would never have these problems. Paradoxically, though, it is the failure to acknowledge and discuss problems that can diminish your learning experience, your performance, and your evaluation by supervisors on site and on campus (Blake & Peterman, 1985).

Moving through this stage often involves taking another look at your expectations, goals, and skills. Although you may have set goals that were reasonable at the time, experience may have shown that some of them are not realistic or the opportunities may have changed. This is also a time to reexamine and perhaps take the necessary steps to bolster your support system.

There may be interpersonal issues between you and your clients, supervisor, or coworkers that are getting in the way. You may need some help clarifying these issues and developing a strategy for resolving them. You will need to consider intrapersonal factors, such as mounting personal problems or unexpected crises in your outside life. There may also be aspects of your personal makeup that are contributing to the problems. For example, it may be that your reactions to some typical features of an internship (such as criticism, authority, or speaking in or before a group) reflect patterns evident throughout your life that are being exacerbated by the internship. There are many strategies for dealing with these intrapersonal issues, and we will explore them in Section Three of this book.

As the issues raised in disillusionment are resolved, morale begins to rise, as does task accomplishment. The task at this stage is to keep working at the issues raised. This is a time when you may be tempted to "freeze the moment" and resist raising any more issues for fear of spoiling the progress you have made. The temptation is normal, but if you give in to it for long, you may find yourself stagnating or even regressing. However, with each new round of confrontation, you will feel more independent, more effective, and more empowered as a learner. You will have a sense of confidence that comes not just from what you have accomplished, and not from denying problems, but from your knowledge that you can grapple with problems effectively.

STAGE 4: COMPETENCE

As your confidence grows, you will forge ahead into a period of excitement and accomplishment. Morale is usually high, as is your sense of investment in your work. Your trust in yourself, your site supervisor, and your co-workers often increases as well. You may find yourself thinking of yourself less as an apprentice and more as a professional. You may even wonder why you are not being paid. As an emerging professional, you have a solid platform from which to expect, or even demand, more from yourself and your placement. You may find that you want more than you are getting from your assignments, your instructor, or your supervisor. Many interns also report that during this time they are better able to appreciate the ethical issues that arise in their placements and are more willing to confront them. These are all positive developments. If taken too far, though, they can lead to perfectionism. You may begin to apply unreasonable standards to those around you, to yourself, or both. Excellence, not perfection, is your goal in this stage.

Another issue that can arise during this time is the stress of juggling your life outside the internship with your increasing commitment to your work. Although you may feel pulls on your time and loyalty throughout your placement, your earlier anxieties and roadblocks may have demanded too much of your attention to think about these conflicts. Now that these earlier crises are past, conflicts between home, school, internship, and friends can surface more easily. This can become overwhelming, especially if you strive for perfection rather than excellence in all these arenas.

STAGE 5: CULMINATION

This stage occurs as your internship approaches its ending date. The end of the internship, coupled with the end of the semester and in some cases with the end of the college experience, can raise some big issues for you. You may experience a variety of feelings as this time approaches. Typically, there is both pride in your achievements and some sadness over the ending of the experience. You may also feel guilty about not having done enough for clients or concern that no one will be as effective with certain clients as you have been (and you may be right). For those of you ending your college career, you may be concerned with continuing your education, employment, or economic survival. Relationships with friends, family members, lovers, and spouses that have been organized around your role as a student have to be reorganized. In any case, there are many good-byes to be said. Good-byes are never easy, and for some people, they are very difficult.

Often, interns find ways to avoid facing and expressing these feelings, particularly the negative ones. Avoidance behaviors may include joking, lateness, or absence. Some interns may devalue the experience—they begin saying it hasn't been all that great or find increasing fault with the placement site and/or clients. Many interns find themselves having a variety of feelings and reactions, some of them conflicting and changing by the hour. This can be very confusing and upsetting. To address the concerns of this stage, you need to focus on your feelings (whatever they may be), have a safe place to express them, and find satisfying ways to say good-bye to clients, staff, supervisors, and in some cases, other interns, both at the site and in an internship seminar on campus.

If you do not pay attention to the concerns of the culmination stage, the internship is, of course, going to end just the same. However, you can be left with an empty or unfinished feeling. In some cases, interns struggling with culmination actually reinvent their placement; they allow their feelings about ending to color their perception of the experience as a whole.

SUMMARY

Now you have a sense of what is ahead in your internship and in this book. Remember that even though these stages may hold true for many or even most interns, especially when viewed from the outside, both the pace with which you move through the stages and the phenomenological experience of being in them will vary a great deal from individual to individual. As you move through the stages of your internship, the chapters in this book will explore the stages in more depth and encourage you to remain focused, as well, on the aspects of yourself you explored in Chapters 2 and 3.

For Further Reflection

1. As you read the description of the stages of an internship, did anything seem remotely familiar? Did the stages remind of you of any other experiences you have had? As you read the stages, did any one in particular stick in your mind or attract your attention? Why?
2. Think about the issues that the stages may have raised for you and those that the theory does not seem to address. Make note of these issues for future reference.

For Further Exploration

Lacoursiere, R. (1980). *The life cycle of groups: Group developmental stage theory.* New York: Human Sciences Press.

Explains Lacoursiere's theory in detail and discusses its application to many kinds of groups.

Schutz, W. (1967). *Joy.* New York: Grove Press.

Another group development theory that has had an effect on our view of internships. Talks a great deal about acceptance, inclusion, and control issues.

Sweitzer, H. F., & King, M. A. (1994). Stages of an internship: An organizing framework. *Human Service Education, 14*(1), 25–38.

Gives more detail on how we used Lacoursiere and Schutz to arrive at our ideas about the stages of an internship.

Sweitzer, H. F., & King, M. A. (1995). The internship seminar: A developmental approach. *National Society for Experiential Education Quarterly, 21*(1), 1, 22–25.

Discusses our general approach to working with interns from a developmental perspective.

References

Blake, B., & Peterman, P. J. (1985). *Social work field instruction: The undergraduate experience.* New York: University Press of America.

Lacoursiere, R. (1980). *The life cycle of groups: Group developmental stage theory.* New York: Human Sciences Press.

Nesbitt, S. (1993). The field experience: Identifying false assumptions. *The LINK (Newsletter of the National Organization for Human Service Education), 14*(3), 1–2.

Sweitzer, H. F., & King, M. A. (1994). Stages of an internship: An organizing framework. *Human Service Education, 14*(1), 25–38.

Sweitzer, H. F., & King, M. A. (1995). The internship seminar: A developmental approach. *National Society for Experiential Education Quarterly, 21*(1), 1, 22–25.

SECTION TWO

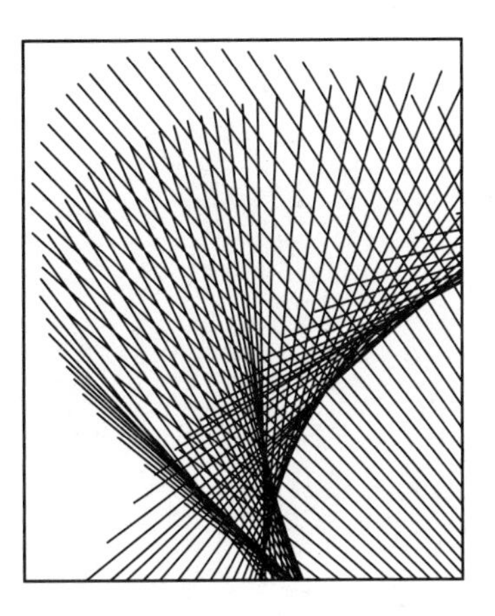

BEGINNING THE JOURNEY

The previous section was the most theoretical, and in some ways the most difficult, of the book. We have given you a lot of theory to understand and think about. Now we get to the business of applying that theory to your internship, moving through the stages of the internship, and helping you meet the challenges of each. The beginning of an internship can be overwhelming emotionally. Experiencing the internship is much different than reading about it or imaging it or even doing role plays and simulations. As you will see, it is also a very exciting time with great intellectual as well as emotional challenges. There is a lot to think about and a lot to prepare for. This section of the book will help you have as smooth a beginning as possible and set the foundation for further progress.

The rest of this section of the book will help you focus on the tasks of the anticipation stage. In Chapter 5, we focus on two concerns that beginning interns often have about themselves: competence and role definition. Next, we invite you to set clear goals and objectives for yourself. In Chapter 6, we help you look at your assumptions about and relationships with your clients, your supervisor, and your co-workers. Finally, Chapter 7 helps you become aware of and address some concerns about the organization as a whole.

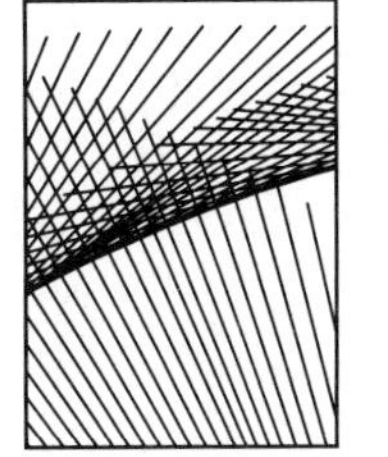

CHAPTER 5

Experiencing the "What Ifs": The Anticipation Stage

> *Many of us seem to have a lot of self-doubt about our capacities and what is manageable for us. It would seem that after all the required academics we would not have these feelings, but apparently not.*
>
> STUDENT JOURNAL ENTRY

Your internship is probably something you have been looking forward to for a while. Some of you have been waiting just a semester or two, others have waited their whole college career, and some have been waiting a lot longer than that! You have heard about it, many of you have worked hard and sacrificed to get here, and now here it is. But as your friends and family ask whether you are excited, and you answer "yes," you may also feel some anxiety creeping in. It may be a little or a lot; it may be visible to your friends, or you may keep it hidden. You may even hide your anxiety from yourself; some interns become aware of it only after it starts to go away (Wilson, 1981). In any case, you are entering an unknown experience, and that's always at least a little scary.

Stop a moment and think about other new experiences in your life. Do you remember your first day of school? Of high school? Of college? How about summer camp or a trip to see relatives you didn't know? An excursion into an unfamiliar neighborhood? You were probably excited about these things, too—and nervous. You have probably heard lots of stories from students who have already completed their internships; that is both a blessing and a curse. It builds your excitement, but you may also hear a little voice inside saying, "Am I going to be as good at this as they are?" You may also have heard some "horror stories" about internships that went sour in one way or another; even though these are usually few and far between, the stories seem to have great staying power.

We call this first stage of an internship the "What if . . . ?" stage. Many interns express their concerns during this stage. They wonder whether they will be able to handle it when __________ happens. Here are some of the "what ifs" we have heard:

WHAT IF:

a client insults me?

a client confronts me?

a client gets physical?

my writing skills aren't good enough?

I can't think of anything to say?

I can't handle all my responsibilities at home and at the internship?

a crisis happens and I don't know what to do?

a client lies to me?

a client falls apart?

I fall apart?

I make a mistake—and something really bad happens as a result?

my clients don't get any better?

clients ask me personal questions?

I say something that offends a client?

I have a client I can't stand?

I lose my temper?

I have to discipline a client who won't listen?

I'm asked a question and don't know the answer?

my supervisor doesn't like me?

the other workers resent or ignore me?

I hate it?

Some of these may seem silly, and no one (so far) has worried about all of them. Your particular anxieties will be shaped by your personality, your knowledge of your placement site, and your past experiences. Every one of the entries on the list, however, is something we have actually heard. You can probably add to them. We encourage our students to share their anxieties with one another, and they are often surprised and relieved to find that they are not alone in their fears and concerns:

> I have come to realize that we are all in the same boat although we are not all physically in the same sites. And, we are all going through the same feelings, worries and anxieties. Knowing this and seeing it in black and white has helped me tremendously.

Every one of those concerns is perfectly normal, and every one can be addressed, or at least the attendant anxiety can be lessened.

THE TASKS AT HAND

Even though you may be a little nervous, you are probably eager to get going. After all, you didn't sign up for this experience to sit in a class, read a book, or write in a journal. You want to start doing the work you signed up for, acquiring new skills, and making a difference. However, at the same time as you start these processes, we are going to encourage you to attend to some other tasks:

- Examine and critique your assumptions.
- Prepare for the development of some key relationships.
- Acknowledge and explore your concerns.
- Set some clear goals and objectives.

These tasks, if done well, will help address the concerns discussed earlier and will form a solid foundation from which to learn and grow. Without that foundation, you may falter sooner and harder.

Interns begin their placements with certain expectations, which are products of assumptions made, correctly or incorrectly, about many aspects of the upcoming or beginning internship (Nesbitt, 1993). As we mentioned in Chapter 4, these assumptions may come from stereotypical portrayals in the media of certain client groups or agencies or from your own experience. Your previous experience, however, does not always predict what the future holds, although it is natural to generalize. You probably have at least some assumptions about clients, supervisors, the field site, and your co-workers. Making these assumptions explicit and subjecting them to critical examination will help you develop the most realistic picture possible of the internship you are beginning.

We said in Chapter 1 that relationships are the context or the medium for learning during the internship. You will be involved in many relationships with clients, supervisors, instructors, co-workers, and other interns during this time, and many of these relationships are new. You are in the beginning stage of those relationships, and a normal concern at this stage is acceptance. No book can tell you how to achieve acceptance with all of these people or even with any one of them. Each of these people is unique; furthermore, it is an interactive process, and half of the interaction is you. All the facets of yourself that you explored in Chapters 2 and 3, and probably more, are part of the acceptance process. A relationship, then, is a process created by you and the other person, and as such is unique. However, there are some things we can help you consider and be aware of as you explore these relationships.

Perhaps the greatest concern for beginning interns is competence (Brust, 1986; Yager & Beck, 1985). Naturally, you want to do a good job and be recognized for it. Furthermore, if you think about all the examples of professionals at work that you have observed, read about, or seen in videos, you have probably seen very few examples of mediocrity; the idea was to show you how effectiveness looks in action (Yager & Beck, 1985). However, a distortion effect can occur. When we go to the movies, we see the scenes that were kept, not the ones that were scrapped or had to be reshot (again and again). Even the most brilliant baseball players fail to hit safely three out of five times!

Yet we never think about the failed attempts; we concentrate on the successes. Similarly, watching those videos and reading those cases make it easy to forget that everyone undergoes a learning process and that everyone makes mistakes. We imagine that we could never do quite as well and recall our own fumbling efforts. We often see only one of the paths to success and imagine we may never find it, while fearing the many paths to failure. So the question is: Can I really do this?

You may have been quite successful in your classes, even in those that focused on skill building. Your previous field experiences may have gone quite well. But interns often tell us that this is the big one, and some feelings are new to them. Some interns, for example, struggle with what has been called the *imposter syndrome* (Clance, 1985). They are vulnerable to believing that whatever success they have achieved is due to good fortune rather than competence. At each step, they fear being exposed for the pretenders they imagine themselves to be. In the case of the internship, they imagine that they did well enough to qualify only through remarkably good luck and that they will surely be found out at their placement site. These are natural concerns, and the feelings about assessment, competence, and initiative that you explored in Chapter 2 will all shape how these concerns manifest for you.

The good news is that there is a cure. The bad news is that, in large measure, the cure is time. You will almost certainly feel more confident and competent as time goes on. In the meantime, if you check with your peers, you will see that you are not alone and that knowledge can have remarkable healing powers.

Another concern for many interns is the nature of their role. As we have explored this issue further with students, it seems that it has two parts. First of all, interns are concerned about how they will be seen by others at the site and how they will come to see themselves. Are you a student? A volunteer? An observer? A researcher? A staff member? In fact, you will probably perform all of these functions, but none of these roles by itself defines who you are at the internship. The role of an intern is unique. Initially, you may spend a good deal of time observing. Later you will be a more full-fledged participant, but it is important to retain some of your observer status (Gordon & McBride, 1996). Because you are there to learn, not just to work, you must reserve some time to engage in and complete the experiential learning cycle we discussed in Chapter 2. You must take time to observe and reflect as well as take action and have experiences.

There may be volunteers at your internship site, and that can cause some confusion for you as well as for the staff. Volunteers and interns do some of the same tasks, but their roles and responsibilities are very different. Volunteers are essentially there to help out. You are there to learn. You will have specific learning objectives the volunteers do not have. As a learner, you should be doing, but you also have the responsibility to reflect, analyze, and critique, which a volunteer does not have. Furthermore, although you may perform some of the same tasks as a volunteer, as an intern you have a responsibility to learn why a particular task is done, why it is done in a particular way, and how it relates to the bigger picture in the organization (Royse, Dhooper, & Rompf, 1996).

Next, interns wonder what it is they will be asked to do. Will they have to dive right in? Will they be given boring, mechanical tasks like copying, filing, answering the phone, or driving? Your learning contract, negotiated with your instructor and supervisor, should help allay some of these fears. However, you should expect to be given some of what is often called *grunt work*. In many agencies, almost everyone pitches in and does some of the drudgery (Royse et al., 1996) You will probably be asked to do your share. You may even be asked to do a little more than your share so that more skilled staff can focus on other kinds of work. It should be clear, though, to everyone concerned exactly who can assign tasks to you. If you feel you are being asked to do too much of this sort of work or being asked by too many different people, you should speak with your instructor and/or your supervisor.

However, as Bruce "Woody" Caine (1998) has pointed out, there is actually quite a bit you can learn from performing these tasks. Often, the learning is not in the task itself, but in the organizational meaning of the task. You get a chance to learn what it really means to be an entry-level worker in that agency. You become familiar with these tasks, which are critical if not always interesting, and how long it takes to complete them (which is easy for people who don't have to do them to forget). You can also learn how much it costs the organization, in material and human resources, to get this work done.

THE LEARNING CONTRACT[1]

At some point, you, your supervisor, and your field instructor will need to negotiate a learning contract. When that happens in the internship is a matter of policy—that of the academic program or that of the field site. Sometimes the contract is developed prior to students beginning their field experiences or shortly thereafter. Other times the learning contract is developed after the students begin their internships, usually within the first 2 weeks of placement. It is created at that time so students can take an active role in designing the learning goals and activities. Making a useful learning contract takes some time, and you may wonder whether it is worth it. In our experience, the process is beneficial to both you and the others involved with your placement.

Setting goals keeps you focused. Think, for a moment, about when you had a day with no particular plans or obligations. Those days can be wonderful—and rare—but you may also have found yourself spending quite a bit of time doing nothing in particular or deciding what to do. That's fine for a day or two of your leisure time but counterproductive in a professional setting. Of course, at your internship, you are not likely to have many days when there are no particular demands. On the contrary, on any given day at your placement, there are usually many competing demands on your time, attention, and energy. You may be approached by clients or co-workers about becoming involved in a project, a conversation, or a group activity. Your supervisor may ask whether you want to work on a particular report, review a policy manual, or visit another agency. Your instructor may ask you to select books or articles to read. All these activities may sound good, but you cannot do them all. Furthermore, if you try to

[1]There are two contracts that are usually involved in an internship. The first is an *internship agreement*, which is a legal document between the campus and the placement site; this contract details the responsibilities of the involved parties and describes in general terms what people will do to carry out the agreement. The second is the *learning contract*, which is a way to make clear to all parties what you want and expect to learn, and in what ways that will happen; this contract includes specific goals and objectives.

spread your energy over too many activities, you will not do a good job on or get the most out of any of them. There may even be a quiet time at the agency when there is nothing concrete demanding your attention, and some interns seem to find that experience disorienting.

A clear sense of goals and objectives will help you use your time well. There seems to be additional power in writing your goals down, at least for many people (Caine, 1998). Writing down goals often increases your commitment to them; at the very least, it saves you from having to memorize them.

Setting goals also keeps others focused. Your learning contract should be negotiated with and agreed to by you, your supervisor, and your instructor. The overall goal at your internship is for you to learn. However, once you get going, it is easier for everyone to see the work you do and to focus on that rather than on what you are trying to learn. Ideally, what you do should flow from what you want to learn, and the goals and objectives describe your learning. You will be specifying learning activities as well, but these may change for one reason or another. If you are clear about the goals, you can choose an alternative activity that will bring you to the same end point.

Finally, a learning contract can be empowering. In many of your learning activities and classes, someone else tells you what you should learn, what activities you will undertake in order to learn, and how you will be evaluated. Sometimes what they choose is a great match for you, your needs, and your style; sometimes it is not. This is an opportunity for you to say what you want. You may be a person who finds it easier to just go along, trusting that the experience will be interesting. While you may be right, you also need to stretch your repertoire to include a more proactive stance on your own behalf, and this is one opportunity to do that.

There are many kinds of learning contracts ranging from simple goal statements that apply to any intern to highly specific and individualized plans (Wilson, 1981). Your program may already have an approach and format for learning contracts. What we will describe is a highly detailed approach. It can be adopted as is or adapted to suit your needs and the needs of your instructor. The process has several steps. It begins with goals, which are general statements, written specifically by and for you, about what you want to accomplish and when. Then we ask you to set objectives, which are more specific statements; they are the building blocks of goals. Next come learning activities, which are the experiences you plan to have that will help you meet your objectives. Finally, assessment entails the ways that you and others will see how well the objectives have been met.

As you approach this activity, remember that the goals, objectives, activities, and assessment procedures that you write down now should not be rigid and unchangeable. You may want to make some changes as you go along. That is fine, as long as the changes are made explicitly and negotiated with the same people with whom you negotiated your original agreement.

Goals

A goal is a general statement about what you want to learn and/or accomplish. It is not a statement about what you are going to do. "I want to lead groups with adolescent alcoholics" is not a goal. "I want to improve my group facilitation skills" and "I want to understand several ways to work with teenage substance abusers" are goal statements. If you are not sure about a particular statement, just ask yourself whether it makes clear what you want to learn. If what you come up with first is a statement about what you want to do, try asking yourself why you want to do this activity. What do you hope to gain? The answer, as illustrated earlier, is a clue to your goal. There are four possible categories for your goals: (a) knowledge, (b) skills, (c) personal growth, and (d) career development. You should consider including some in each.

Knowledge goals describe new information you want to learn, such as:

- Understand how the criminal justice system in this state operates.
- Know what services are available for the physically challenged in this city.
- Be more familiar with common physical problems experienced by people in their seventies.

Skill goals describe things you want to learn to do. Skills can be related to knowledge, of course. You may learn the rules of a behavior management system (knowledge), but you still need to learn to use that system with clients, and that is a skill. Here are some other examples:

- Be able to intervene in a crisis at the shelter.
- Learn to write a case report.
- Improve oral presentation skills.

Personal growth and development goals involve the ways in which you hope to grow and change as a person. Most interns describe a good deal of personal growth, some of which they never anticipated, but here we are asking you to think about what you might like to focus on, understand, and change about yourself. Here are some examples:

- Understand why I have trouble being patient sometimes.
- Learn to be more assertive.
- See whether I have the compassion it takes to do this work.
- Become more open-minded and less judgmental about poor people.

Finally, *career development goals* focus on ways that your experience can help you clarify your career plans. For example:

- Find out whether I have the skills and stamina for child protective work.
- Learn what an administrative job is really like.
- See whether I enjoy working with the elderly.

An important issue in designing goals is the level of specificity. You do not want your goals to be too narrow and specific; that is the job of your objectives, which come later. At the same time, if goals are too broad, they are not going to be easy to monitor or to achieve. Furthermore, they don't tell others specifically what you want. "Increasing my counseling effectiveness" is too broad. "Improving my ability to give positive praise" is too narrow. "Improving my one-to-one counseling skills with adolescents" is specific enough to serve as a guide and broad enough to encompass a variety of objectives.

Objectives

Objectives are more specific statements that form the components of your goals. They should be written, whenever possible, in observable terms. If you have specified an objective in observable terms, then anyone watching you can tell whether it had been met. So, for example, you might say, "I want to better appreciate my client's cultural identity." Appreciation is not observable. Ask yourself how someone would know when you had become more appreciative. The answer might be, "I could describe each client's cultural identity and some of its impact on their behavior." Now that is observable, and it is a great objective.

Try to write objectives for every goal. If you cannot, perhaps your goals are too specific. In that case, ask yourself what larger end is served by a particular goal. Then, you can make the goal an objective. For example, suppose you had a goal that said, "I want to learn more about demonstrating empathy." You might struggle to come up with objectives. However, if you ask what larger purpose is being served here, the answer might be that you want to learn more about Rogerian counseling or that you want to connect with a variety of clients. In that case, the statement you originally had as a goal becomes an objective, and the more encompassing statement becomes the goal.

Learning Activities

Now that you have set your objectives, it is time to consider what kinds of activities and experiences will help you achieve each one (if you chose not to specify objectives, you can still do this with your goals). The first step is to take each objective and think about the obstacles to achievement. Consider, for example, the objective discussed earlier, "Describe each client's cultural identity and its impact on their behavior." What are some obstacles you might face?

- You don't know the racial composition of the client base.
- You don't know much about Hispanic culture or Inuit culture.
- You are nervous around homosexuals.
- You're not sure how to ask a client about his or her background.

The next step is to identify people, experiences, and material resources to help you overcome these obstacles. The obstacles could be addressed through a combination of reading, observation, practice, and consultation with your supervisor or co-workers. In your learning contract, you would specify the readings, observations, and other resources.

Now you are ready to identify some specific activities. This step, if not the earlier ones, should be done in consultation with your instructor and supervisor. They may have ideas for you, or you may have ideas they never considered. Also, something you propose may not be possible. For example, suppose you decide you want to read client files, but your supervisor tells you that files are confidential unless you are working with that particular client. If you and the supervisor are clear about the objective being met by that activity, you can come up with another activity to meet the objective. You may also want to consider attaching a timetable to the activities so that everyone is clear about when they are going to happen.

As you suggest learning activities or respond to the suggestions of others, remember your exploration of learning style in Chapter 2. Choose activities that take advantage of your learning strengths, but also some that stretch you and challenge you to expand your repertoire. If you prefer concrete modes of getting information, then you will want to do more activities and fewer readings. That is fine, if your supervisor agrees, but consider accepting, or even asking for, some readings as well.

Assessment

How will you and others know that your objectives have been met? Your learning activities specified the things that you thought would help you, but you don't know for sure how much they helped, if at all. Furthermore, chances are that you engaged in other activities of an unplanned nature or that helped you with more than one objective. A systematic look at how you did and what you learned will help you accomplish your goals. An assessment is not a grade or a rating. An assessment is a chance to see where your strengths are and where you need to improve. You should think about how you want your progress to be assessed. Examples of assessment activities include:

- Videotaping yourself
- Surveying clients or co-workers
- Being observed and given feedback
- Writing a case study for review

Try to be creative, and remember that these activities are there to help you. If you assess your learning as you go along, you have a chance to focus your energy on improving in specific areas. You also get the chance to feel the pride of accomplishment. That is important, especially if you happen to be in a setting where the clients tend to be rather unappreciative of you and other staff members. Furthermore, you will be undergoing formal assessment and evaluation at least once during the internship (this topic is discussed in more detail in the next chapter). If you have been gathering information on your progress all along, you can be a more active participant in this process.

Levels of Learning

There are many ways to carry out the internship you and your supervisor design. In many instances, supervisors have supervised interns in the past and have even been interns themselves. However, some supervisors are new to the role and will be learning along with you how best to develop your learning experience.

Part of the process of developing a learning contract is paying attention to different levels of learning that need to take place along the way (King, Spencer, & Tower, 1996). Learning takes time, and your activities must be organized so as to build on your attitudes, skills, and knowledge as they develop and change over time. We find it useful to think in terms of three sequential levels of learning: (a) orientation, (b) apprenticeship, and (c) mastery.

The first level, orientation, is a time when you become acquainted with the work of the internship. Days at the site are marked by lots of observing, reading, shadowing, and inquiring. These are far from aimless activities. You need to have goals during this time because you are building a foundation for a successful internship (Feldman & Weitz, 1990). The second level, apprenticeship, is an intensive teaching and learning period. The third level is a period of mastery during which you will be engaged in your own work, structuring and organizing your schedule so as to meet time lines and agency goals. You are more self-directed, and your need to understand subtle aspects of the work and yourself is heightened. You will also need less direction and more support from your supervisor.

Since you are now in the first level of learning, it may be difficult to project what your goals will be later, especially in the mastery level. Do the best you can to think through what you will want to learn and do during all three levels, and remember to come back and look at your learning contract frequently. It should not be rigid, but instead should be a clear and flexible guide to your internship.

SUMMARY

You should expect some anxiety as you approach your internship. We hope this chapter has helped you clarify what some of your anxieties are and where they come from. We have also addressed the process of setting learning goals and activities, which should help alleviate some of the anxiety that comes from uncertainty. Many of the concerns discussed earlier, though, have to do with the many relationships you are beginning to form. In the next chapter, we turn to these relationships and the ways you can get the most from them.

For Further Reflection

1. Look at the "what ifs" listed in your book. How many of them seem familiar to you? Can you add anything to the list?
2. Make a list of both your anticipations—the things that give you feelings of hope and excitement—and your anxieties. If you have any thoughts about where your anxieties come from (past experience, personal issues, etc.), please include them.

   Optional: Leave your name off this exercise. Bring it to class and put it, with those of your classmates, in a box. Have your instructor read them aloud or make copies for everyone. Note areas of similarity and difference.
3. As interns approach their internship, they often use metaphors to describe what they think the experience will be like, such as a roller coaster ride, a glider flight, or a trust walk. Try to think of a metaphor that captures the way you are thinking, feeling, and behaving as you approach your internship. Be sure to explain as well as describe it.
4. Have you experienced feelings of fraudulence in your role as an intern? Do you ever feel like an imposter and fear being discovered as less competent than you should be and appear to be? Discuss your experiences with this phenomenon.

For Further Exploration

Wilson, S. J. (1981). *Field instruction: Techniques for supervisors.* New York: Free Press.
Excellent, thorough coverage of learning contracts. Reviews various types of contracts and assessments.

References

Brust, P. L. (1986). Student burnout: The clinical instructor can spot it and manage it. *Clinical Management in Physical Therapy,* 6(3), 18–21.

Caine, B. (1994, Winter). What can I learn from doing gruntwork? N.S.E.E. *Quarterly,* pp. 6–7, 22–23.

Caine, B. T. (Ed.). (1998). *Understanding organizations: A ClassPak workbook.* Nashville, TN: Vanderbilt University Copy Center.

Clance, P. R. (1985). *The imposter syndrome.* Atlanta: Peachtree Publishers.

Feldman, D. C., & Weitz, B. A. (1990). Summer interns: Factors contributing to positive developmental experiences. *Journal of Abnormal Behavior,* 37, 267–280.

Gordon, G. R., & McBride, R. (1996). *Criminal justice internships: Theory into practice* (3rd ed.). Cincinnati, OH: Anderson Publishing.

King, M. A., Spencer, R., & Tower, C. C. (1996). *Human services program manual.* Unpublished manuscript, Fitchburg State College.

Nesbitt, S. (1993). The field experience: Identifying false assumptions. *The LINK (Newsletter of the National Organization for Human Service Education),* 14(3), 1–2.

Royse, D., Dhooper, S. S., & Rompf, E. L. (1996). *Field instruction: A guide for social work students* (2nd ed.). New York: Longman.

Wilson, S. J. (1981). *Field instruction: Techniques for supervisors.* New York: Free Press.

Yager, G. G., & Beck, T. D. (1985). Beginning practicum: It only hurt until I laughed. *Counselor Education and Supervision,* 25(2), 149–157.

CHAPTER 6

Getting to Know the People

The interpersonal work is the critical aspect of any internship—graduate or undergraduate. It is also the most difficult for most people

STUDENT JOURNAL ENTRY

Regardless of how bright or motivated you are, or how much you know, it is your ability to form and maintain effective relationships with a wide range of people that will make you successful in your internship as well as in your career in human service. In this chapter, we help you look at some of those important relationships—those with your clients (if you have them), your supervisor, and your co-workers—and examine your expectations and assumptions. We will also call to your attention some issues you may not expect; they will be easier to deal with if they are not surprises. Finally, we deal with acceptance concerns, which are a normal part of the beginning of any relationship.

CLIENTS

Working in a human service agency or other organization, it is easy to become preoccupied with paperwork, policies and procedures, and office politics. As an intern, you will have all these concerns to deal with plus the additional concern of being evaluated by at least two people. However, as J. Robert Russo reminds us in his book, *Serving and Surviving As a Human Service Worker* (1993), the people your organization helps are the reason for your job. Many internships involve providing direct service to people in need; those are your clients. If you are having your first experience with providing direct service, it is natural that many of your thoughts and much excitement and anxiety are directed at the clients. Even if your placement does not involve direct service, you may still want to skim this section in order to understand the broader context of your agency's work. We have found that concerns about clients generally fall into two categories: the nature of the client population (which involves your assumptions) and your relationship with clients.

The Nature of the Population

As you begin your work, you no doubt have some expectations and assumptions about the clients you will meet. You probably have some knowledge of your clients as a group from the placement process and your first couple of days at the site. You know in general terms who the agency does and does not serve and some of the needs the agency can and cannot meet. Beyond this factual information, though, it is important to be aware of the image you have of your prospective clients. You are bound to have one, and it is not all based on facts.

Imagine your clients for a moment. Each one will be somewhat different, but you can describe them within certain parameters. Client populations can be described as homogeneous or heterogeneous with regard to various characteristics. The more homogeneous a population is, the more alike its members are; the more heterogeneous the population, the more diversity is found within it. So, for example, a high school has a population that is relatively homogeneous in age, but may be heterogeneous with regard to characteristics such as race, social class, and intellectual acuity. Think about your clients' backgrounds, their reasons for coming to the agency, their personalities, life histories, and typical behaviors. Think about their race, ethnicity, social class, religion, and sexual orientation. How heterogeneous do you imagine the population to be? In our experience, interns often imagine that the clients will be very similar to one another, or they go to the other extreme and see each one as totally different, missing some of the important commonalities among them.

Now think about how similar and/or different the clients are from you. What do you have in common with them? What aspects of their lives and experience are totally unfamiliar? Here we often find that interns imagine they have almost nothing in common with their clients. That is, of course, not entirely true, but it is easy to feel that way. On the other hand, interns who have struggled with the same problem that brings clients to the agency (such as alcoholism or domestic violence) may assume that they know just how it is for their clients. A little thought, and some reflection on classes you have taken, will tell you that's not true either, but it is another tempting assumption. Are you falling into either of these traps?

You have been thinking about the image you have of your prospective clients. Now think about the source of that image. Probably very few of you have had extensive and varied experience with the client population you will be encountering. In spite, or perhaps because, of their inexperience, people make unconscious generalizations; they form stereotypes. The word *stereotype* has some pretty negative connotations, and you may be reluctant to consider that you have some. However, whenever you make a judgment about someone based on little or no information, you engage in stereotyping. Try imagining two people. One is a slim, slightly pale man in a tweed jacket wearing wire-rimmed glasses and carrying a beat up briefcase. The other is a tall, heavy man with a large belly, long hair, and a big beard dressed in a T-shirt, sunglasses, dirty jeans, black boots, and a black leather vest with a Harley Davidson insignia on it. In spite of yourself, are you making assumptions about what each of them does for a living? About how educated they are? Their personalities? Which one is more likely to have read existential philosophy? To have been in a barroom brawl? It seems to be a human tendency to generalize, and the fewer people in any group we actually know, the more we are likely to generalize from the few that we do know or have read about.

Returning to your client population, your assumptions can come from several places. Perhaps you have met a person with the same needs or heard one speak. Perhaps you have been such a client yourself or know someone who has. The media are another powerful source of images and assumptions. There are scholarly books and documentaries on various client groups, but a more powerful source of images is the mainstream media, especially television. Think, for example, about how many mentally challenged adults you have seen depicted on prime-time television. Were any of them leading anything like normal lives? How many of them had committed a crime? Couple the images of this population portrayed in the media with your own lack of direct experience and you can begin to see where your image and stereotypes may originate.

As we said in Chapter 2, the first step in getting beyond stereotypes and assumptions is to admit you have them. Check with other interns; they have them, too. Theirs may not be the same as yours, but they have them. The next step is to gather as much factual information as you can and hold your assumptions up to the light of objectivity. At the beginning of this section, we asked you several questions about your clients and asked what you were assuming. Now may be a good time to go back and try to find factual answers to those questions. Don't be discouraged if the answers don't come easily; you have held onto some of your assumptions for a long time.

Another common concern is being successful with your client population. How would you define success with this population? Be as specific as you can. What attitudes, skills, and knowledge do you think it takes to be successful with them? This may be a good time to talk with others at your placement site. Their input will help you obtain realistic answers to these questions as well.

Your Relationship With Clients

As interns think about the kind of relationships they want to have with clients, several concerns often arise. One of the most common is how clients will react to you. In talking with students who were going to perform community service with the homeless, Ostrow (1995) reported that one of their major anxieties was how they, as relatively affluent and fortunate young people, would be perceived by the homeless. You may be wondering, "What kind of reception am I going to get from these people? Will they respect me? Will they listen to me? Or will they just write me off?" The theme in these concerns is acceptance, which is a crucial part of the foundation for any relationship. You have probably read about how important it is for you to accept the clients, but they need to accept you, too. You need to find ways to have them accept you, and that's not always easy.

You are probably also concerned about your reaction to the clients and about how you will handle certain situations. Here are a few of these concerns that have been found in interns and beginning helpers, drawn from our experience and the work of

Cherniss (1980), M. S. Corey and Corey (1998), and Russo (1993). How might you react to clients who:

- lie to you?
- manipulate you to get something they want but cannot have?
- are never satisfied with what you have to give and always seem to need more?
- become verbally abusive and physically threatening?
- blame everyone else for their problems?
- are sullen and give, at most, one word answers or responses?
- ask again and again for suggestions and then reject every one?
- refuse to see their behavior as a problem?
- make it clear they don't like you?
- refuse to work with you?

All of these things can and do happen, although not as often as you may fear. Even though they happen infrequently, it is normal for them to be concerns; they are important challenges in developing relationships with clients. The rest of this section will discuss some of these client issues and concerns in more depth.

Acceptance

Think for a moment about what the word *acceptance* means to you. Don't recite a definition; try to put it in your own words. Ask a peer to do the same. You could probably have a debate about exactly what it means. Most people, though, know what it feels like when they are accepted and when they are not. Think about how acceptance feels to you as well as what it means.

Acceptance can look very different with different client populations. Some clients show their acceptance simply by being willing to talk to you; until they accept you, they just ignore you. Other client groups may show their acceptance by including you in their conversations, confiding in you, considering your suggestions, following your directions, or accepting the limits you set. How do you imagine you will know whether your clients have accepted you? Be as concrete and specific as you can.

Getting clients to accept you can be a real challenge. In some cases, clients may be particularly wary of interns. Some respond better to volunteers, reasoning that they are there because they want to be, as opposed to interns, who are fulfilling a requirement for school (Royse, Dhooper, & Rompf, 1996). In addition, interns generally do not stay very long. Some may be there for one academic year, and some may even be hired, but many are gone after one semester. Many clients have had unfortunate experiences with people leaving them or letting them down, and they are reluctant to invest in a relationship that will end soon. One of our students, who was working at a group home for adolescents, reported that one of the residents was quite hostile and rebuffed all attempts at contact. On further inquiry, the intern learned that this resident had gotten very close to an intern the prior semester and, although she knew the intern would leave, was devastated when it happened.

Part of developing acceptance with clients is finding some common ground. A frequent concern for interns is their belief that they have nothing in common with their clients. A young woman interning in a shelter for single women and their children said:

> I don't have any kids. I have no idea what it's like to be a single parent, or to grow up with one parent. I have no idea how to survive on the streets or what it's like to be abused by a husband. What do I have to offer these women? Why should they listen to me?

Here is another thought from an intern working with adult drug addicts:

> I couldn't survive a minute in their world, and they know it. I didn't grow up in the city or in the street culture. I haven't been in a fight since grade school. And I'm going to give these guys advice?

These are legitimate concerns, and in fact, clients probably won't accept these interns until they figure out some way to bridge the gap. You can't manufacture experience you haven't had, and pretending you know how it is will only make things worse. However, common ground does not always mean common experience, and showing respect for your clients' experiences is one way to gain their acceptance.

At the other extreme are interns who have had some similar experiences. They come from the same background or have struggled with the same problem as their clients. This common experience can be a help in establishing acceptance, but it can also be a hindrance. You may have had some of the same things happen to you, but assuming or saying you know how your clients feel can be alienating for a client whose experience of the same problem was quite different from yours.

The answer to these dilemmas is not simple, and it will take some time to find your way. We have found that interns who really listen—to clients and to themselves—gain acceptance the quickest. If you stay open to the experiences, background, thoughts, and feelings of clients and to your own thoughts, feelings, and current and past struggles, you will eventually find common ground on which to meet.

OTHER ISSUES

There are a number of issues that seem to arise frequently as interns launch their relationships with clients. Many of these issues are related to acceptance; they are also related to one another. Being aware of them will help you prepare and not be shocked when they appear.

One particularly challenging issue concerns authority. In Chapter 3, we asked you to reconsider your feelings about being in a position of power or authority, which is perhaps a new experience. However, you have certainly been in a position where you were subject to authority and so have your clients. Authority is often an important issue in forming a relationship with them. Shulman (1983) has pointed out that clients tend to perceive you as an authority figure. This is especially true if you or your agency has some kind of legal authority over clients, such as the authority a benefits worker has to deny benefits or a probation officer has to surrender a client to the court. However, even when the authority is not explicit, you should be prepared for clients who see and react to you as an authority figure (Shulman, 1983).

Very often, clients are in a "one-down" position. They need something they cannot get for themselves. It could be something tangible, such as food or shelter, or something less concrete, such as control over a substance abuse problem or help in understanding some bureaucratic procedure. You and your agency have what they need; you are holding the cards. Think about when you have been in a similar situation. How did it feel? Your clients will bring their fears about past experiences with authority to their relationship with you, and those factors will shape how they respond.

One way clients try to reduce their one-down feelings is to assess your background and experience, often with a goal of finding a flaw or an equalizer. We refer to this phenomenon as *credentialing,* and it takes place in nearly every internship. Sometimes clients will literally ask about your credentials, as when they ask about your education and training or experience with clients. Other times the credentialing is more subtle, such as, "Do you have children?" or "Have you ever been arrested?" It may be as simple as asking about your age. Sometimes clients are just trying to get to know you, but you may be surprised at the persistence with which these questions are asked. You may also be surprised by their reaction when they find the information they are looking for, and you may feel dismissed. Knowing that these assessments are a normal part of building a relationship will help you decide how to respond.

Clients will also often test your limits. They want to see where your personal boundaries are and whether and how you enforce agency rules. Because you are an intern, they may be genuinely unsure of your role and what you can offer them (Shulman, 1983). Some clients have experienced many workers, not to mention other people in authority, and have been treated in many different ways. They need to know what to expect from you, and although they may not like it when you set a limit, it ultimately helps them trust you. Other reasons for testing you may include a need for recognition and attention or an attempt to gain status with their peers (Shulman, 1983).

You may think of this phenomenon as occurring more with children, as they try repeatedly to pick up something they have been told to leave alone or try to poke or bite you. You may also associate testing with people who are in an involuntary situation, such as juveniles in a detention center or runaway shelter, who refuse to do chores or curse in front of you to see what you are going to do. However, other kinds of clients can test you, too. An elderly client can press you to stay longer than you are able; a client at a soup kitchen may try to go through the line twice; a parent may ask you to stay late and watch the children until they can be picked up.

Interns are sometimes tested even more because they are initially seen as "not real staff" (Gordon & McBride, 1996). These behaviors can be exasperating, especially if you are not expecting them. Try not to imagine that really effective workers never have these challenges. Of course they do. They may have learned to handle them a bit more smoothly and quickly—and you will, too, someday. Taking these challenges personally only makes it harder to meet them effectively. Remember that you once did this kind of testing, and you may still do it occasionally (Russo, 1993). Think back for a moment on your behavior with substitute teachers or baby-sitters. Perhaps you have even tested some of your college instructors in this way.

Another important issue in developing client relationships, and a frequent concern for interns, is self-disclosure. How much about yourself should you reveal to clients?

There are actually two kinds of self-disclosure at issue. The first is personal information. It is natural for clients to want to get to know you, and they may ask you questions about your life and relationships. Some of these questions will seem quite comfortable and easy to answer. Others may not. You may feel that the questions are too personal or that they concern matters you would rather keep private. You may feel that certain information is inappropriate for certain clients. Given the earlier discussion of credentialing, you may also wonder about the motives behind the question.

The second kind of self-disclosure is more immediate. You will no doubt have opinions about and emotional reactions to things your clients say or events at the site and wonder whether it is appropriate to share your thoughts with clients. The question, "What should I share with clients?" is not possible to answer and, in fact, is not a helpful question. It fails to consider the wide variety of clients, situations, and goals that form the context of your work. It is better to ask a "three-dimensional" question (Hunt & Sullivan, 1974), such as: What sort of self-disclosure is appropriate with which clients and for what purpose? If you are working with a battered woman, for example, it may be appropriate to share some of your relationship struggles as a way of establishing an empathic connection; clients sometimes think their counselors have no problems of their own. However, if you are working with a heterosexual teen of the opposite sex, such disclosure is probably inappropriate, as it can blur an important boundary and create confusion about your intentions. Asking three-dimensional questions will help you find answers that work for you and your clients.

In Chapter 2, we stressed the importance of being aware of your values. In their book *Issues and Ethics in the Helping Professions*, G. Corey, Corey, and Callanan (1998) stress the importance of values awareness in relationships with clients. There will be times when you are dealing with a client whose values are very different from yours, and that can be perplexing and stressful for both of you. For example, suppose honesty and straightforwardness are strong values. For you, to be dishonest is wrong. Of course, you are dishonest occasionally, but you know it is wrong when you do it. However, you may be dealing with a client who has different values. In her experience, being honest, especially with human service workers and other authorities, means being taken advantage of by a service delivery system she perceives as unfair. For example, in some states, a woman on welfare will have her benefits reduced if she is married, even if her husband is not employed. So, she lies to you.

You may also encounter clients whose values about sexuality are different from yours. For example, you may have an adolescent client who is sexually promiscuous and thinks it is just fine. Furthermore, he uses no birth control and says a pregnancy would not be his responsibility. In the area of family values, you may be working with a client who thinks that a marriage must stay together at all costs, but you are in favor of divorce in some cases. Or you may be talking with a client who thinks that unmarried couples should not have children.

One of the challenging aspects of situations like these is deciding how to respond. G. Corey et al. (1998) make a distinction between exposing and imposing your values on clients. You must decide whether to expose differences by telling clients about them and also whether to attempt to influence clients to change their values (which is imposing).

One important way to form an effective initial relationship with clients is to try and understand the values underlying their behavior. That way, it will seem less like a willful transgression, although you may still object to (or in some settings, even punish) the behavior.

Another important factor in your initial (and continuing) relationship with your clients is their cultural identity and yours. Working with clients with a significantly different set of cultural and subcultural influences requires careful attention to them and to yourself. If you are a member of some dominant subgroups (please review the definitions of dominant and subordinate in Chapter 2), your clients may be especially cautious around you. Furthermore, approaches and assumptions that work well with clients from a cultural background similar to yours may not work very well with a different population. Finally, both your and your clients' stage of dominant or subordinate identity development will have an impact on the relationship. Many of you have probably studied this topic already, and this book is not the place for a lengthy discussion of multicultural issues. However, we provide some references for further reading at the end of the chapter.

YOUR REACTIONS TO CLIENTS

For the last several pages, we have been discussing common concerns about your clients' reactions to you. Prior to that, we discussed other concerns that interns have about how they are going to react to their clients. You may be wondering how you will respond to clients, or you may have had an emotional reaction to one or more of the hypothetical situations posed. Even if this did not happen, you will surely have some emotional reactions to your clients, and they will probably be of many different varieties. The most important thing you can do is try to understand your reactions. Remember, just as clients bring their past experiences and emotional tendencies to the relationship, so do you.

Now is a good time to review this section and note your reactions to the various clients and situations you read about. Then go back to Chapters 2 and 3 and reconsider some of the things you discovered or affirmed about yourself. Do any of those things help explain your reactions? For example, think about the reaction patterns you may have. If you have trouble confronting others, then clients who challenge you overtly will be especially difficult for you. Knowing this will help you avoid assigning all the blame to the client. If your psychosocial identity, as discussed in Chapter 2, includes a shaky sense of competence, then clients who are not making progress may actually anger you more than they should. Finally, your cultural identity may help explain some of the reactions you are having as well. Consider this quote from a student journal:

> Some clients evoked feelings of mistrust and prejudice, as well as feelings of sorrow. I am most shameful of the feelings of prejudice.

This quote shows an intern who is willing to look to him- or herself as well as to the client to discover the source of this reaction. The willingness to admit prejudice is especially impressive. In time, this intern may move past feelings of shame and work on ways to overcome the prejudice and mistrust. Admitting these feelings, instead of pretending not to have them, is the first step.

This section on clients may have raised more questions than answers for you. That is the point. We are trying to make you aware of and prepared for some of the challenges in the early stages of working with clients. We have provided you with some good material for reflection and for discussion with your peers, your instructor, and your supervisor. However, the exact shape and pace of your experience with clients are things we cannot know.

SUPERVISORS

Certainly, one of the most important relationships at your internship is the one you have with your supervisor.[1] You will spend a great deal of time with these people, and they will be responsible at least in part for your evaluation. Your relationship with your supervisor is a tremendous source of learning about the work, about yourself, and about the relationship itself (Bogo, 1993; Borders & Leddick, 1987).

Three definitions of supervision help capture the complexity of this role and relationship. Gordon and McBride (1996) define it as "a process where two people . . . meet for the primary purpose of enhancing the personal and professional development of the student" (p. 32). Collins (1993), on the other hand, says that supervision is "a dynamic relationship of unequal power, which is characterized by intermittent closeness and distance" (p. 122). Still others see the role of supervisor as embracing both of these definitions. Moses and Hardin (1978) perhaps describe it best as going beyond instruction and developing skills and knowledge to promote self-actualization. Supervision should provide an extending experience that blends personal knowledge and personal qualities.

Your supervisors can and should help you learn a great deal. However, they also indisputably have power over you, and that has a real effect on the relationship. Interns report a variety of feelings about and relationships with supervisors. There is no one right way. In fact, a good relationship with a supervisor can take many forms. Consider these two examples from student journals:

> I have strong convictions and so does my supervisor. We do not always see eye to eye. I admire people who can stand up for their convictions.

> My supervisor seems to know exactly what my needs are as an intern. She has never assigned me something I was not ready to do. I think the most important aspect of [an] internship is having good communication with the supervisor. And we have it.

This section will help you examine the beginnings of your unique relationship with your supervisors and consider some important aspects of that relationship.

Because the relationship is so important in so many ways, you would be unusual if you did not have some concerns as you approach it (Costa, 1994). Perhaps primary

[1]In some cases, interns have more than one person functioning as an on-site supervisor. Your instructor on campus may also be fulfilling some of the functions of a supervisor. In this section, though, we are assuming that you have one on-site supervisor.

among those concerns is acceptance. Interns often wonder whether supervisors will like them. More important, they wonder whether their supervisors will understand them and accept their weaknesses. As in any new relationship, they wonder whether they are going to get along well.

Self-disclosure is a concern in this relationship as well, although not in the same way as it is with clients. You may wonder how much about yourself you want to share or are expected to divulge and how many of your feelings and reactions you ought to reveal as you go through the internship. Since you do not yet know these people, you may wonder how they will react to your disclosures; some interns, especially in counseling settings, worry that their supervisors will take the opportunity to analyze them and expose personal and professional weaknesses (Wilson, 1981). You may also be concerned about being assessed and evaluated. Your supervisors' perceptions of you are important in and of themselves, but the added prospect of a formal evaluation adds extra weight to those perceptions. Your own issues and personality will determine how important or intense any of these concerns are, and you may have some others we have not mentioned.

Again, the cure for some of this is time. As you get to know your supervisor, you may or may not like what you find, but at least it will be known. And a trusting, comfortable relationship takes time to develop. However, clarity can also help, and that is the focus of this section. This is not a section about supervision theory. Rather, it is designed to help you become a more intelligent consumer of supervision. Looking at yourself as a consumer of supervision, with rights and responsibilities, is a much more empowering stance that looking at yourself as a passive recipient or even a victim. Being an intelligent consumer involves learning who your supervisor is as a person and a professional and examining the match between what you know about yourself and your needs with what your supervisor has to offer. It also involves clarifying some important procedures in the supervisory process, including the method and means through which you will be evaluated.

The Supervisor As Person and Professional

In our experience, interns often bring certain preconceptions about supervisors to the internship (see also McClam & Puckett, 1991). They often think of their supervisors in a two-dimensional way, as if interns were the supervisors' only responsibility. Interns also tend to assume that anyone who is a supervisor must have a great store of experience and expertise. This may or may not be true. Your relationship with your supervisor will be smoother if you let go of some of these preconceptions.

Your supervisor is a person with concerns, foibles, strengths, and weaknesses just like you. You will get to know some of these over time. Right now, though, you can find out more about who your supervisor is in the organization. Many supervisors agree to take on an intern because they see it as a way to contribute to their profession or perhaps to return the favor that a supervisor did for them when they were students. There are some more tangible benefits to working with interns as well. For example, some supervisors have sought out interns because of the positive effects the supervisors observed on the morale, commitment, and proficiency of their staff as a result of working with interns (King & Peterson, 1997).

However, remember that your supervisor probably has a supervisor as well, or maybe several of them. He or she undoubtedly has a job description. Agreeing to supervise an intern is an investment of time, and that time is lost for other agency activities (Royse et al., 1996). If the investment pays off and the intern does a good job, then everyone benefits, and the intern may even be able to take some of the work load off the supervisor. If it is a difficult internship, on the other hand, and it requires a great deal more time and energy than anticipated, the return for the agency may not be as great. These concerns are probably on the mind of your supervisor.

Try not to make assumptions about your supervisor's level of experience, which may not be a great deal more than yours. However, experience is not always the best indicator of quality in a supervisor. Just as brilliant scholars may not make good teachers, because students' struggles with the material are incomprehensible to them, so it is that experts in your field may not be the best supervisors. They may have forgotten what it is like to be new to the field, and the gap between your skill levels may be so great that they cannot function as effective role models for you—the expertise seems unattainable. So, a less experienced person is not necessarily a problem. Regardless of her or his level of expertise, your supervisor has a different, more objective vantage point from which to view your experience and your struggles.

Your supervisor probably has some special skills in the organization and may have some special talents as a supervisor as well. However, a supervisor fulfills many roles (Borders & Leddick, 1987; McCarthy, DeBell, Kanuha, & McLeod, 1988; Royse et al., 1996), and few are equally good at all of them. Your supervisor is in a teaching role when helping you set goals and objectives, helping you learn new skills, and being sensitive to your particular learning style. Supervisors also function as evaluators. As such, they tell you not just how well you did, but what more you can learn. Supervisors use counseling skills to establish a supportive relationship with interns and to help them through difficult emotional times and decisions. The role of consultant is evident when a supervisor helps an intern assess a problem or situation and generate alternative courses of action. Finally, supervisors are in the role of sponsor when they take an active interest in the career of interns, both at the placement and beyond (Kanter, 1977; Speizer, 1981).

SUPERVISORY STYLE

In all these roles—teacher, evaluator, counselor, sponsor and consultant—your supervisor has a unique style. Just as learning style is multifaceted, so is supervisory style; many different dimensions are involved. Some aspects of your supervisor's style may be the result of careful study and deliberate choice. Others are the result of attempts to replicate the way she or he learned best or to emulate a favorite supervisor. Still others are matters of personal preference. Some aspects of a supervisor's style may not even be deliberate; she or he may be quite unaware of it or assume that everyone does things in the same way.

No one style of supervision is good for everybody. You may have heard great things about your supervisor, yet be puzzled when the relationship doesn't seem to be working. It may be that you are a person who is better suited to a different style. Furthermore, many authors have discussed a developmental approach to supervision, suggesting that the needs of supervisees vary according to the stage they are in (Galassi

& Trent, 1987; Hersey & Blanchard, 1982; Loganbill, Hardy, & Delworth, 1982). An important step in developing a relationship with your supervisor is to learn more about some of the key components of your supervisor's style: theoretical orientation; the level of instrumental, as opposed to expressive, approaches used; the balance between support and direction; and whether a collaborative or hierarchical approach is taken.

The work of an individual, and sometimes of an entire organization, is usually guided by a particular theoretical orientation to the work. There are, for example, many different theories about teaching, counseling, management, and sales. It is important that you know what theory or theories are the most important to your supervisor, and how tolerant of other styles your supervisor is. If, for example, you favor a humanistic or psychodynamic approach to counseling and your supervisor believes in a cognitive behavioral approach, you will need to know to what extent you are expected to use that style with your clients.

The terms *expressive* and *instrumental* describe two general approaches to management or supervision (Russo, 1993). Supervisors who use expressive approaches are more people oriented; the primary concern is for people and relationships. Supervisors who use this style often need to be liked and appreciated, and they cultivate friendships with those they supervise. A supervisor with a more instrumental orientation, on the other hand, is primarily concerned with productivity and task accomplishment. Please note that this is perfectly possible even when the business of the organization is people. A supervisor at a social services agency can be concerned primarily with serving the maximum number of clients and achieving measurable results. For these supervisors, respect is important in the supervisory relationship, but not necessarily affection and friendship. They are able to handle conflict and hostility from supervisees effectively.

M. S. Corey and Corey (1998) discuss a related issue when they compare supervisors who use a more confrontational approach with those who are more supportive. Confrontation is not necessarily an angry encounter. A confrontational supervisor will spend less time offering support and more time helping you see what you could have done differently. For example, if you have just had a difficult encounter with a client, your supervisor could spend time listening to you talk about how that felt and offer some reassurance and empathy. On the other hand, she or he may ask a series of pointed questions designed to help you see where you misjudged the situation and/or used an ineffective intervention. This latter technique does not necessarily signify a lack of caring about you; it is merely a different way of caring and a different perception of what you need.

Hersey and Blanchard (1982) have identified two dimensions of supervision, which can be combined in four different ways. Direction is the first dimension, which involves giving clear, specific directions, close supervision, and frequent feedback. Support, the other dimension, is a nondirective approach marked by listening, dialogue, and high levels of emotional support. So, if your supervisor tells you to write a report and says, "There are lots of good ways to do this, and I know you are nervous, but you can handle it. Please ask all the questions you need to and let me know when you are done," this is a high level of support. A directive approach might tell you exactly what the report should look like and what it should contain, with frequent deadlines for the submission of sections and the final report.

These two dimensions can be combined in four ways: high support/low direction; high support/high direction; low support/high direction; and low support/low direction. You may think that high support and high direction would be the best all the time. Of course, it is frightening to be given a task and have no idea how it should be done. However, too much direction over a long period of time can discourage independence and deny you the opportunities to try something and learn from it. Sometimes you need to take risks, and even make mistakes, and that is difficult when you are very closely supervised. Similarly, everyone needs some support and encouragement, but too much can keep you from seeing where you need to improve. It can also make you dependent on your supervisor in a different way; there may be times when you need to function without considerable external support. Hersey and Blanchard (1982) suggest the most appropriate combination of support and direction depends on the developmental stage of the supervisee; a new employee needs a different approach than a veteran.

A final dimension of supervisory style to consider is collaborative versus hierarchical approaches (Costa, 1994). A collaborative approach emphasizes mutual dialogue between supervisor and supervisee and encourages divergent thinking. In a hierarchical approach, however, the supervisor serves as a communicator of expert knowledge. Expertise flows in one direction, as do the questions, except when the supervisee needs clarification on something.

Keep in mind that each of the terms used here to describe supervisory style defines one end of a continuum. There is plenty of room in the middle. As you get to know your supervisor, you can locate her or him on each of these continua. Remember, though, that supervision is a relationship, and so far we have only covered one side.

YOUR STYLE

As you read through the last section, you probably found yourself identifying your preferences among the styles discussed. If not, take time now to review the section. As you do, remember that your reactions and preferences are influenced by a variety of factors. Your learning style was discussed in Chapter 2. You may learn best by reading, by observation, or by direct experience. Your supervisor may have a different style, assigning readings, for example, when you are longing to deal with real people and situations. Another factor to consider is your motivational style (Borders & Leddick, 1987). Some people respond best to warmth and praise and wilt under blunt critiques, whereas others may become impatient with too much warmth and praise and prefer a relatively impersonal, objective critique.

In Chapter 2, we also encouraged you to consider some of your issues and patterns. It is especially important that you consider any patterns or unfinished business you may have with issues of authority. We already asked you to think about the challenge of being in authority. Here we are asking about negative experiences or reactions you have had to people with authority over you. Some people have a hard time accepting authority. Some go to the other extreme and have difficulty challenging it, even when a challenge is warranted. Sometimes older students have a separate set of challenges in supervision (Royse et al., 1996). Perhaps they have managerial experience in another field. Perhaps they spent a long time rising through the ranks to get there. If so, it may be hard for them to accept being in a subordinate position in supervision, especially if there are some aspects of their supervision that they do not like or that leave them feeling disempowered. Your psychosocial identity may also be important to remember in your relationship with your supervisor. For example, your sense of competence—one central component of that identity—may be challenged by supervision, and your sense of initiative will be important when you are given projects of your own.

Another dimension of your style is as a separate or connected knower. This style may lead you to prefer abstract analysis over in-depth case study, or vice versa. Twohey and Volker (1993) have pointed out that both feelings and principles have a place in supervision. If you are having a difficult time with a client, for example, it is important to examine the theoretical and ethical principles that are relevant to the case. However, you may have feelings that are not easily reduced to a theoretical quandary or a clash of ethical principles. You may be more concerned about the client and the potential disruption of your relationship with him or her. This aspect of your experience needs to be attended to as well.

MATCHES AND MISMATCHES

If you—or your supervisor—are expecting a certain kind of response and get a very different one, it can be very disconcerting. If you are looking for direction and get support, or expecting collaboration and get hierarchy, for example, it can be confusing and upsetting. A supervisor expecting a theoretical analysis who instead gets a long look at your emotional reactions may feel that you either did not understand or are not capable of answering the question. These moments are never easy, but they are easier if they are understood as a mismatch of styles. Your supervisor is not necessarily insensitive to your needs any more than you are stubborn or inept. You may each be doing exactly what you think is appropriate and required.

Misunderstandings can be avoided by taking a proactive stance toward your supervision. Most supervisors do not mind being asked about their style. They may or may not be familiar with the specific theories discussed here, but you can find a way to ask about these aspects of style without using the particular jargon described in this chapter. Your supervisor may also be familiar with and use other theories about supervisory styles. If so, you can learn both about that theory and about your supervisor when discussing style. You may also want to consider discussing your style with your supervisor, including your preferences and needs. These needs do not have to be demands. You are merely considering your style, along with your supervisor's style, and looking for potential areas of match and mismatch.

The Process of Supervision

Part of what creates anxiety for anyone is the unknown. Each supervisor and placement site is different, of course, but in our experience, there are some common features of the supervisory process, and you will want to be clear with your supervisor about each of them.

GATHERING INFORMATION

Supervisors need information about interns in order to know how they are progressing and developing. This information guides supervisors in their conferences with you and in the evaluation process. There are several ways for supervisors to gather this information. The approaches all have strengths and weaknesses, and your supervisor may use more than one of them. It should be fairly easy for you to find out which of these approaches will be used.

One common approach is live supervision, which means that your supervisor either works alongside you or observes as you work. The observation is sometimes done behind a one-way mirror, so that neither you nor your clients can see the supervisor, although you know you are being observed (and the client may be told as well). There is no substitute for live supervision; it is the only way the supervisor can see you work firsthand. As such, it is a rich source of information for the supervisor and hence of useful feedback for you. However, live supervision is often not possible. It can be intrusive on the relationships with a client. It can also make interns nervous, and more important, having your supervisor right there and available for consultation may encourage you to be overly dependent on that help.

Another somewhat "live" approach is the use of audio- or videotaping. You make a tape of yourself and go over it with your supervisor. Some supervisors will go over the tapes privately first, and some may ask you to prepare an analysis of the tape as well. While no one we know enjoys being taped, it can be of enormous value. Watching or listening to yourself is an excellent way to notice voice patterns, body language, patterns of movements, and other aspects of your performance that you would never be aware of otherwise.

Self-report is another common approach to data collection. Your supervisor will ask you to report, orally or in writing, on your progress with clients or projects. This approach encourages you to reflect on your own performance and to take responsibility for critiquing your own work, which is something you are going to have to learn eventually. However, it will never uncover the kind of unconscious patterns and blind spots that live supervision and taping can reveal. There is also a natural human tendency to describe yourself in the best possible light, especially when you want to make a good impression (Borders & Leddick, 1987).

Peer report is a less common approach. In cases where your supervisor cannot observe you directly but other staff members can, the supervisor may request written or verbal reports from these staff members. In the case of written reports, you may or may not be able to see them. It is important to discuss your placement's policy on gathering information on your performance as early as you can.

CONFERENCES

You should expect to have regular meetings with your supervisor throughout the internship. This is a time for you to report on your progress, ask and answer questions, and get feedback. Lots of supervisors will also meet with you spontaneously if there is something important to discuss, and that is fine. However, those meetings allow you no time to prepare, reflect on your experience, and think about issues and questions you may have (Royse et al., 1996). We strongly recommend that some of your meetings be at regularly specified times. Find out from your supervisor how frequent and how long these meetings will be and schedule them ahead of time.

Another important consideration is the structure of the meetings (Borders & Leddick, 1987; Leddick & Dye, 1987). Some supervisors use a highly structured format. They ask the questions, and they ask at least some of the same ones each time. They may also have a structured way that they want you to present your cases or your progress on projects. Other supervisors are very loose in structuring the session and respond only to what comes up in the conversation. Some supervisors come into conferences with a set agenda. Others let you ask the questions and raise the issues. Some, of course, do both.

Regardless of the structure, it is important that you reserve some time to plan for these conferences (Royse et al., 1996). Remember the learning cycle described in Chapter 2. At most placements, there is so much to do that there is a temptation just to work, work, and work some more. Conferences force time for reflection, which is a critical part of completing the cycle, and preparing for them does as well. Planning for conferences can also empower you and give you more control and influence over the session. You should set aside time to think about questions, observations, and problems you want to discuss. You should also be ready to report on your progress with clients and projects.

YOUR REACTION TO SUPERVISION

Even under the best of circumstances, the process of supervision often produces at least some anxiety. In your internship, you are being asked to try out new skills and approaches. In your supervision conferences, you are often encouraged to discuss your feelings and reactions. A supervisor that feels you are running into some unresolved personal issues is probably going to let you know that. Under those circumstances, most of us experience two seemingly contradictory emotions. On the one hand, we are there to learn and grow, and supervision is a tremendous opportunity for that. On the other hand, we do not always want to see ourselves clearly or change our ways of doing things. We both desire to learn and change, yet resist learning and changing (Borders & Leddick, 1987).

These conflicting emotions sometimes manifest themselves in behaviors that help you avoid really looking at yourself and your progress (Borders & Leddick, 1987; Costa, 1994). They include being overly enthusiastic, avoiding certain topics, and going off on tangents. You may also become forgetful, especially about projects, assignments, or issues that make you nervous. Some interns become argumentative, taking issue with every point the supervisor makes. They may or may not do so out loud, but privately and to their friends they dispute every criticism. Others go to the opposite extreme and agree with the supervisor even when that is not how they feel. If you find yourself exhibiting any of these behaviors, think about the feelings and concerns that may be behind them. The point here is not that any single instance of one of these behaviors indicates that you are avoiding an issue; however, when they become patterns, you should look very carefully at them, discuss them with your instructor, and find a way to work through them.

The Evaluation Process

At some point in your internship, you will probably have a formal evaluation. At this time, the supervisor lets you know how you are doing and have done overall. In our experience, most internship programs request two evaluations: one at the midpoint and one at the end. It is natural to be apprehensive about the prospect of reading or hearing an evaluation. But try to remember that the evaluation is another opportunity for growth, empowerment, dialogue, and even assertiveness. Above all, it is an opportunity to learn. Again, the more you know about the process, the less mysterious and threatening it will be, and the more you can direct your emotional energy at making the most of these opportunities.

You can take the initiative to find out several things about how you will be evaluated. The structure of your evaluation is an obvious concern. Find out how often you will be evaluated and whether it will be oral, written, or both. Gather information on the specific format. Some written evaluations use scales, where you are given a number from 1 to 5 or 1 to 10, to indicate how well you have met various criteria. Some evaluations let the supervisor write in a more narrative form. Still others use both formats.

It is also important to know what standards are being used in your evaluation (Wilson, 1981). Some supervisors will compare you to other interns, either current ones or those from past semesters. If you are among the best, you receive high ratings. Others evaluate interns individually, assigning ratings based on how much they have grown over the course of the semester. Still others have a standard set of expectations and criteria that you must meet to get a high rating. If the supervisor is using numerical ratings, find out what they mean. What do you have to accomplish to earn a 3?

We knew one student whose supervisor had nothing but praise for her at their weekly conferences. Yet the midsemester evaluation contained ratings no higher than 3 on a scale of 1 to 5. The supervisor explained that he never gives anyone a 5, because no one is perfect, and if he gave 4s at midsemester, there would be no room for improvement. To him, the 3s were an excellent midsemester evaluation, but to the intern, they felt like an average rating, much like receiving a C on a paper.

You will also want to know the function of the evaluations. Who can see them and for what purpose? Are the midsemester evaluations recorded and counted as part of your grade? What portion of your internship grade is determined by the supervisor's evaluation? You should also clarify when and under what circumstances you will see your evaluation. Many supervisors will make the written evaluation part of a conference. They may discuss the evaluation before filling out any forms, or they may go over the written evaluation with you. Regardless of the specific arrangements, you should see the evaluation before it goes to anyone else and have an opportunity to review it and ask questions.

CO-WORKERS

Although your supervisor may be your most important colleague at the placement site, you may actually spend more time with other staff members. Sometimes interns report that a particular staff member becomes an informal supervisor and that they learn more from and feel more supported by that person than their assigned supervisor, whom they may only see once a week. Whatever the relative importance of co-workers to you, they are a great opportunity for learning and a potential source of support. They may be mentors, sponsors, and role models offering you knowledge and skills, an insider's and veteran's perspective on the agency and its work, and valuable feedback on your performance. Just as with your clients and supervisor, you probably have an image of your co-workers as well, and it needs to be tested. And just as with clients and supervisors, another major concern of the anticipation stage is acceptance from co-workers.

Expectations

You may not have thought much about it, but you probably have an image of what your co-workers will be like. What level of education do you think is needed for working at your site? What level of education do you think most of the staff has? Earlier, we asked you what attitudes, skills, and knowledge were necessary to work successfully with the clients at your site. If you skipped the section on clients, take a moment now and think about what it takes to be successful at your site, not as an intern but as a full-time professional.

Most people are guided by some standards or principles in their work. Your agency has a set of rules and policies and may even have a code of ethical behavior. You may have studied ethics and other aspects of professional behavior in your classes as well. To what extent do you think your co-workers are aware of these issues and standards? To what extent do you think they adhere to them? What do you imagine happens to them if they violate those standards?

We ask these questions because in our experience interns encounter all sorts of co-workers, some of whom behave very differently than they were expecting, as the following student journal entries make clear:

> During recreation period, I am out there trying to make contact with the kids, and the other staff members just sit and talk with each other—and they're the ones getting paid!

> In private, they make jokes about the residents. They have derogatory labels for almost every one of them.

> It seems that my main value to some of them is as a gofer. They give me jobs they don't want, so they can take a break.

> I was amazed to find out that several of the staff here are brand new in the field. With my previous field experiences and practicum, I have more experience than they do.

> Some of them don't have degrees in human services or psychology. Some of them don't have degrees at all!

You can, and should, spend some time getting accurate information about your co-workers. Find out what the hiring process is at your placement. What level of education is in fact required? What other requirements and preferences does the agency have? How many staff people leave the agency in an average year? What do people have to do to get promoted? You may be quite surprised by the answers to these questions.

Acceptance

Most interns are concerned about whether, when, how, and on what terms they will be accepted by the staff at their placement site. As you ponder this question, there are a number of issues to consider. Obviously, different individuals will react to you differently. Based on what you know now, who are the people at your site whose acceptance is most important to you? Remember, too, that acceptance is not the same as being liked. Most of us want to be liked, but the staff members do not have to like you in order to accept you. If you need every staff person to like you, you may find yourself taking action directed primarily at achieving that liking rather than at what is best for your agency or your clients. Finally, remember that acceptance is a two-way street. One challenge for you is to accept your co-workers for who and what they are. Again, you do not have to like all of them or even respect everything they do. In your classes, you may have studied and practiced how to be accepting of clients whom you may not like or approve of. Your co-workers are not clients, and they are accountable to standards that your clients are not, but they deserve the same accepting, nonjudgmental treatment that you extend to your clients.

Depending on the size of the organization hosting your placement, you will probably get a range of reactions from the staff, from warmth to indifference and even to hostility. Some interns report feeling like guests, some like employees, and still others like intruders. Try putting yourself in the place of the staff members. It is another busy day at work, and they are meeting an intern whom they may or may not have known is coming. They may be wondering what your presence means for them, what you want from them, exactly why you are there, and whether they can count on you.

The staff's reactions to you depend on many factors, including their personalities, past experiences with interns, understanding of what an intern is (or lack thereof), and their relationship with your supervisor (Gordon & McBride, 1996). Some of them may even resent you. Energetic new workers can be threatening to those who are disenchanted with or exhausted by their work. A new intern's ideas can be threatening, or some staff members may fear for their jobs. Furthermore, agencies sometimes use interns and volunteers as a way to stretch their staff and meet their needs without hiring additional people (Suelzle & Borzak, 1981). This practice may violate labor laws or ethical policies and often results in a lack of continuity within the agency, as well as resentment from front-line staff members who have to deal with the results.

Even staff members who have no experience with interns may have heard stories about interns from your school or other schools. As we mentioned earlier, when an intern works out well, it is productive and exciting for everyone involved, but if not, it can be very taxing on staff time and energy and can be disruptive to the work.

Gordon and MacBride (1996) have noticed that interns often struggle with feelings of intrusion and marginality. If you feel like you really don't belong at the placement, that you don't fit in and can't contribute much, you are struggling with feeling like an intruder. On the other hand, if you are feeling like the staff doesn't want you there and doesn't see that you can be of much help, then you are struggling with marginality. Of course, as the following student journal entries show, both can occur at once:

> I feel like my co-workers treat me like an adolescent if I don't present myself properly.

> They tell inside jokes and don't clue me in. One of them in particular just looks at me, as if I don't belong there and she is waiting for me to prove it.

In fact, you may feel like an intruder at first, and you may be treated like one. It may take some time to prove yourself to the staff. Over time, these feelings and reactions usually dissipate, but they are troubling while they are present.

You may also find yourself surprised by the variety of approaches to the job taken by your colleagues. Some may be very different than what you expected. You may be disappointed in them, and you may have a right to be. However, understanding a little bit about what can happen to a person over time may help you understand what you are seeing.

First of all, it is possible that some staff members are experiencing burnout. Burnout is the result of persistent job-related stress. Its effects include physical and emotional exhaustion, rendering the worker depleted and drained; depersonalization of work and clients, so that the worker feels detached and even callous; and reduced personal accomplishment (Maslach, 1982). The exhaustion is not just physical; it is emotional and even spiritual. The person feels irritable and displays negative attitudes toward clients. Productivity declines and the person often feels isolated and withdrawn. Perhaps you have had firsthand experience with burnout, encountering it in teachers, school counselors, or staff at other field sites.

You will find people who are burned out in every profession, but human service workers seem especially vulnerable. The kinds of people who select human services as a career are typically concerned with individuals and their problems, attuned to human suffering, and anxious to make a difference. These very qualities make them vulnerable to working too many extra hours or to putting their clients' needs ahead of their own (Cherniss, 1980; M. S. Corey & Corey, 1998; Schram & Mandell, 1997). There are also many features of human service work that can cause stress. Clients are often not appreciative of your efforts, change often comes slowly and sometimes is not seen for years, many agencies are understaffed and underfunded, and neither the pay nor the prestige is equivalent to other professions.

Christine Maslach, who has written extensively on burnout in the helping professions, emphasizes that burnout is a progressive process (Maslach, 1982). Most burned out workers didn't start out that way, nor did they get that way quickly or abruptly. Most interns swear they'll never be "like that." But burnout sets in slowly, usually without your knowledge. One of the major misconceptions that new workers have is that burnout only happens to "bad" workers, something they are not and will never be (Sweitzer, 1995). Burnout is relatively easy to address during the early stages, but once a person is deep into burnout, it is very hard to recover. It is easy for interns to feel judgmental about and superior to workers who are in the midst of this struggle. Of course, these workers can have a negative effect on the work at the agency in many ways, but they deserve respect and assistance. You probably cannot assist them, but you can respect them.

Burnout is just one possible outcome of a career in the helping professions. Many people avoid this outcome and have productive, satisfying careers. Nevertheless, over the course of many years in the field, there are adjustment to be made. There are several ways to cope with the ongoing demands—both physical and emotional—of the work and the day-to day strains and frustrations that are as much a part of the work as the joys and satisfactions. Russo (1993) has described several patterns of adjustment that are found in experienced workers. They fall into three categories: (a) workers who identify with the clients, (b) those who identify with their co-workers, and (c) those who identify with the organization. Russo emphasizes that these are not static categories; people often move from one to another and back over the course of their careers. It is probably far too early in your own career to determine which category fits you, but knowledge of them may help you make sense of what you are seeing at your placement.

Workers who identify primarily with clients can be further divided into four subcategories. *Reformers* tend to be impatient with anything that they believe interferes with their ability to serve the clients. They often neglect paperwork and will try to change the organization's policies and procedures to better meet the needs of the client (i.e., people before paper). In our experience, this is a very common stance for interns, although they do not actively try to change the placement site. *Innovators* still promote change, but are more patient with and understanding of the change process. They listen, they ask questions, and they work with others to try to find the best way to achieve change. *Victims*, on the other hand, are frustrated by systems they see as inadequate and even harmful. They see themselves as the protectors of the clients and are likely to battle the administration openly, sometimes enlisting clients (or interns) in their struggle. Finally, *plodders* identify with the clients and may have some of the same concerns and frustrations as some of the other types described, but they seem to have given up on change. They work quietly with their clients, doing the best job they can. They often have their own way of doing things, and they work without making waves, and do not try to influence others or the organization itself to change.

Workers who identify primarily with their co-workers also care about clients, but in addition, they feel a strong allegiance to their profession as a whole. They may be active in unions and/or professional organizations and look to these groups, as opposed to their particular workplace, as their primary guides. Some relatively new workers are attracted to this stance because they are unsure of their own skills and knowledge. They follow the rules of the professional organization rigidly. Other workers in this category are more sure of themselves and regularly consult their union or professional organization for guidance, but consider these groups as one of several sources of wisdom.

Finally, workers who identify with the organization look to the organization and its policies and procedures as their primary source of guidance. Even though those rules may sometimes work against the needs of a particular client or be in violation of standards or ethical codes issued by professional organizations, these workers believe that following the rules will do the most good for the most people in the long run. Some may be hiding behind the rules so they do not have to think hard or take risks. Others have adopted this stance after careful thought and reflection on their experience. Russo also points out that some of them are conflict avoiders. They realize that the needs of clients, the organization, and the profession can sometimes conflict, but they want to resolve those conflicts quickly, and adherence to the rules is one way to accomplish that.

Becoming comfortable with a new group of people takes time. How much time depends on the situation you are in, the people that you work with, and the person that you are. Ideally, you will enter your relationships with co-workers with a clear set of expectations and prepared for some of the challenges of acceptance.

SUMMARY

Relationships form the primary context for your learning at the internship. As such, beginning them well is an important part of meeting the challenges of the anticipation stage. These are probably relationships you will remember for all of your professional life. Your first group of clients—those you work well with and those you do not—is usually unforgettable. If you prepare well for them by questioning your assumptions and gathering as much factual information as possible, you will have a better chance of success. Similarly, examining your stereotypes about your supervisor and co-workers will improve your chances of a smooth entry and should reduce your anxiety as well. Acceptance is an issue in all these relationships, as it is in so many areas of life. Thinking about the issues raised in this chapter, and about those raised in Chapter 2, will let you think about and approach the acceptance process from both sides of the relationship.

Although people are of critical importance, they do not tell the whole story of an internship. To complete the picture of your internship and your journey through the anticipation stage, you must consider your placement site as an organization. That is the subject of the next chapter.

For Further Reflection

CLIENTS

1. What did you know about your clients when you started? What did you know about that population in general (e.g., juvenile delinquents or homeless)? Where did that knowledge come from? Did you have experience with the population before?
2. Did you have any impressions or stereotypes about this population? Had the media affected your impressions? Where else did your impressions come from?
3. Now that you have gotten to know the clients a little bit, how have your initial impressions changed? Or have they remained the same?
4. How would you characterize the clients at your agency? In what ways are they similar to each other (race, gender, ethnicity, personality, social class, etc.)? In what ways are they different?
5. In what ways are your clients different from you? What do you have in common with them?
6. As you think about working with these clients, what are the most important things to you about how they respond to you? About how they treat you? About how they feel about you?
7. Have your clients challenged your credentials? If so, how? If not, how do you think they might do so?
8. How easy do you think it will be for you to set limits with this population? What will be the issues around which you will need to set them? If the prospect seems difficult, how much of that difficulty is because of what you know about the clients and how much is because of what you know about yourself?

9. What aspects of yourself are you willing to share with your clients? What aspects are you unwilling to share and/or seem inappropriate to share with this particular population? How will you respond if you are asked about these areas?
10. What other kinds of challenges are you expecting from clients? Include general as well as specific challenges (i.e., those that pertain to one particular client). Why do you think these issues and behaviors will challenge you? Think about yourself. Is there something about you that makes these issues and behaviors especially troublesome?
11. If you have had other internships or field experiences, compare your current group of clients and your reactions to them with previous client groups.

YOUR SUPERVISOR

1. What is your supervisor's primary theoretical orientation?
2. Characterize your supervisor on the following continua:
   a. instrumental vs. expressive
   b. support vs. direction
   c. collaborative vs. hierarchical
3. In what ways is your supervisor's style well matched to yours? Are there any areas of mismatch?
4. Supervisors have different ways of finding out how you are doing. These methods include direct supervision, self-report, and peer review. Which method or methods does your supervisor use? Which method or methods would you prefer?
5. Does your supervisor follow a structured format? If not, who decides what will be discussed? If you don't have any particular questions or concerns, will your supervisor bring up topics?
6. How are you going to be evaluated by your supervisor? What form will be used? What criteria will you be judged by? What happens to the evaluation after it is written?
7. Here is another exercise for those of you who have had other internships:
   a. Compare and contrast the supervision experiences you have had. You should include, but not necessarily limit yourself to, items 1 through 6.
   b. Compare your reactions to your supervision experiences. It's not always the case that one is better or worse, but they are always different and they are bound to affect you differently.
   c. What are the things you need from a supervisor? If you were "shopping" for one, what would you be looking for?
   d. Do your positive and negative reactions to supervision fit into any patterns in your life? For example, are some of the things that make you uncomfortable in supervision also things that make you uncomfortable in other areas of life? Are you happy with these patterns?

YOUR CO-WORKERS

1. How would you characterize your co-workers? What is their level of skill? Experience? Professionalism? Are they similar or different from what you expected?

2. How do your co-workers seem to be responding to you? How do you feel about that response?
3. What do you hope your co-workers will be able to give you and do for you?
4. Do you recognize any of the patterns of adjustment discussed in this chapter (see pp. 99–100)?
5. Are any of your co-workers showing signs of burnout?

For Further Exploration

Axelson, J. A. (1999). *Counseling and development in a multicultural society* (3rd ed.). Pacific Grove, CA: Brooks/Cole.

An excellent text on multicultural issues.

Corey, G., Corey, M. S., & Callanan, P. (1998). *Issues and ethics in the helping professions* (5th ed.). Pacific Grove, CA: Brooks/Cole.

Excellent section on the ethics of imposing values on clients.

Corey, M. S., & Corey, G. (1998). *Becoming a helper* (3rd ed.). Pacific Grove, CA: Brooks/Cole.

Helpful sections on client issues and supervision.

Gordon, G. R., & McBride, R. B. (1996). *Criminal justice internships: Theory into practice.* Cincinnati, OH: Anderson Publishing.

Excellent discussions of issues relating to clients, supervisors, and co-workers. Obviously, this book is aimed at interns in a particular kind of setting, but it has many applications outside criminal justice.

Hersey, P., & Blanchard, K. (1982). *Management of organizational behavior: Utilizing human resources* (4th ed.). Upper Saddle River, NJ: Prentice Hall.

Situational leadership theory is a very popular approach to management and supervision. Discusses combinations of support and direction and when each combination may be most helpful.

Pedersen, P. (Ed.). (1985). *Handbook of cross cultural counseling and therapy.* Westport, CT: Greenwood Press.

Another good resource on multicultural issues. Triad model is helpful.

Pedersen, P. (1988). *A handbook for developing multicultural awareness.* Alexandria, VA: American Association for Counseling and Development.

A very "hands-on," short book with lots of exercises to help you.

Royse, D., Dhooper, S. S., & Rompf, E. L. (1996). *Field instruction: A guide for social work students* (3rd ed.). New York: Longman.

Excellent section on supervision.

Russo, J. R. (1993). *Serving and surviving as a human service worker* (2nd ed.). Prospect Heights, IL: Waveland Press.

Useful perspectives on supervision and co-workers. Especially thorough treatment of patterns of adaptation found in veteran workers.

Schutz, W. (1967). *Joy*. New York: Grove Press.

In his first stage of group development—inclusion—Schutz talks a great deal about acceptance concerns and the various ways of handling them.

Shulman, L. (1983). *Teaching the helping skills: A field instructor's guide*. Itasca, IL: F. E. Peacock.

Especially helpful regarding authority issues.

Sue, D. W. (1981). *Counseling the culturally different*. New York: Wiley.

Classic text by one of the leaders in the field

Wilson, S. J. (1981). *Field instruction: Techniques for supervisors*. New York: Free Press.

Useful discussions of supervision and evaluation.

References

Bogo, M. (1993). The student/field instructor relationship: The critical factor in field education. *Clinical Supervisor, 11*(2), 23–36.

Borders, L. D., & Leddick, G. R. (1987). *Handbook of counseling supervision*. Alexandria, VA: Association for Counselor Education and Supervision.

Cherniss, C. (1980). *Professional burnout in human service organizations*. New York: Praeger.

Collins, P. (1993). The interpersonal vicissitudes of mentorship: An exploratory study of the field supervisor-student relationship. *Clinical Supervisor, 11*(1), 121–136.

Corey, G., Corey, M. S., & Callanan, P. (1998). *Issues and ethics in the helping professions* (5th ed.). Pacific Grove, CA: Brooks/Cole.

Corey, M. S., & Corey, G. (1998). *Becoming a helper* (3rd ed.). Pacific Grove, CA: Brooks/Cole.

Costa, L. (1994). Reducing anxiety in live supervision. *Counselor Education and Supervision, 34*(1), 30–40.

Galassi, J. P., & Trent, P. J. (1987). A conceptual framework for evaluating supervisor effectiveness. *Counselor Education and Supervision, 26*, 260–269.

Gordon, G. R., & McBride, R. B. (1996). *Criminal justice internships: Theory into practice*. Cincinnati, OH: Anderson Publishing.

Hersey, P., & Blanchard, K. (1982). *Management of organizational behavior: Utilizing human resources* (4th ed.). Upper Saddle River, NJ: Prentice Hall.

Hunt, D., & Sullivan, E. (1974). *Between psychology and education*. Hinsdale, IL: Dryden.

Kanter, R. M. (1977). *Men and women of the corporation*. New York: Basic Books.

King, M. A., & Peterson, P. (1997). *Working with interns: Management's hidden resource.* Workshop presented at the annual meeting of the American Probation and Parole Association, Boston.

Leddick, G. R., & Dye, H. A. (1987). Effective supervision as portrayed by trainee expectations and preferences. *Counselor Education and Supervision, 27*(2), 139–154.

Loganbill, C., Hardy, E., & Delworth, U. (1982). Supervision: A conceptual model. *Counseling Psychologist, 10*(1), 3–42.

Maslach, C. (1982). *Burnout: The cost of caring.* Upper Saddle River, NJ: Prentice Hall.

McCarthy, P., DeBell, C., Kanuha, V., & McLeod, J. (1988). Myths of supervision: Identifying the gaps between theory and practice. *Counselor Education and Supervision, 28*(1), 22–28.

McClam, T., & Puckett, K. S. (1991). Pre-field human service majors' ideas about supervisors. *Human Service Education, 11*(1), 23–30.

Moses, H. A., & Hardin, J. T. (1978). A relationship approach to counselor supervision in agency settings. In J. D. Boyd (Ed.), *Counselor supervision* (pp. 437–512). Indianapolis, IN: Accelerated Development.

Ostrow, J. (1995). Self-consciousness and social position: On college students changing their minds about the homeless. *Journal of Qualitative Sociology, 18*(3), 357–375.

Royse, D., Dhooper, S. S., & Rompf, E. L. (1996). *Field instruction: A guide for social work students* (3rd ed.). New York: Longman.

Russo, J. R. (1993). *Serving and surviving as a human service worker* (2nd ed.). Prospect Heights, IL: Waveland Press.

Schram, B., & Mandell, B. R. (1997). *An introduction to human services* (3rd ed.). New York: Macmillan.

Shulman, L. (1983). *Teaching the helping skills: A field instructor's guide.* Itasca, IL: F. E. Peacock.

Speizer, J. J. (1981). Role models, mentors and sponsors: The elusive concepts. *Signs: Journal of Women, Culture and Society, 6*(4), 692–712.

Suelzle, M., & Borzak, L. (1981). Stages of fieldwork. In L. Borzak (Ed.), *Field study: A source book for experiential learning* (pp. 136–150). Beverly Hills, CA: Sage Publications.

Sweitzer, H. F. (1995). Burnout: Avoiding the trap. In H. Harris & D. Maloney (Eds.), *Human services: Contemporary issues and trends* (pp. 215–230). Boston: Allyn & Bacon.

Twohey, D., & Volker, J. (1993). Listening for the voices of care and justice in counselor supervision. *Counselor Education and Supervision, 32*(3), 189–197.

Wilson, S. J. (1981). *Field instruction: Techniques for supervisors.* New York: Free Press.

CHAPTER 7

Getting to Know the Placement Site

I learned the rules by observing my surroundings and conversations.
STUDENT JOURNAL ENTRY

DON'T SKIP THIS CHAPTER!

You may be looking at the title of this chapter and wondering what is going on. After all, you have already read about and considered the clients, your supervisor, and your co-workers. You understand your role, you have a learning contract, and you have read about the stages of an internship. What else is there? Although you do know quite a bit about the placement site, you only have part of the picture. Part of your orientation and adjustment to your internship is learning how the agency or company you work for operates and why it operates that way. This knowledge will help you make sense of your experience there, although it may not seem like it right away.

In addition, if you are not prepared for the organizational dynamics and issues discussed in this chapter, you could be in for some unpleasant surprises. Suppose, for example, that you are working in a geriatric facility. Your supervisor, who meets with you once a week, is not always on the floor with you, but a shift supervisor is there most of the time that you are. This person seems unusually hostile and appears to resent the time you spend with your supervisor. She also occasionally contradicts something your supervisor has told you. While this situation would not be easy under any circumstances, it will be less mysterious if you understand that your supervisor was recently promoted to that job and the shift supervisor was passed over, even though she has been there longer. Whether you know it or not, organizational dynamics are bound to affect you, your clients, your colleagues, and your supervisor.

YOUR PLACEMENT SITE AS A SYSTEM

We are going to encourage you to look at your placement site through the lens of systems theory, which you may have studied in some of your classes. Even if you understand each person in the system as an individual, to understand the system you must understand the way everyone interacts; the whole, in this case, is greater than the sum of its parts (Berger & Federico, 1985). A system is a group of people with a common purpose who are interconnected such that no one person's actions or reactions can be fully understood without also understanding the influence on that person of everyone else in the system (Egan & Cowan, 1979).

Systems can be analyzed internally or externally (Berger & Federico, 1985). Internal analysis involves studying the inner workings and components of the system and the way human and material resources are arranged and expended. However, it is important to realize that all systems are hierarchical. This does not mean that they use a hierarchical authority structure; some systems do not. But each system is part of a larger system, and most can be broken down into smaller systems. A family services center, for example, can be broken down into the various programs it runs. However, it is also part of a system of service providers in the city or town in which it operates. An external analysis examines the relationships between a system and other related systems.

There are formal aspects of systems, which tend to be written down and are relatively easy to find and learn. However, an equally important part of systems is the informal or unwritten dynamics. These are harder to see, especially for someone beginning an internship, but they are important, and you can learn to look for them over time. We have chosen a few concepts from systems and organizational theory that we think are particularly relevant to your work as an intern. We will not be covering all, or even most, of the major concepts in systems or organizational theory, and you may want to investigate them further, or even take a course in organizational behavior.

We begin with some of the components of an internal analysis. One important characteristic of any system is its organization, and this aspect of a system has several components, including rules, roles, communication, decision making, and evaluation. We also discuss the importance of history, goals, values, and philosophy, as well as the financial resources of the organization and how they are acquired and allocated. Each of these components of your placement site has both formal and informal aspects.

THE FORMAL ORGANIZATION

For most beginning interns, the formal aspects of the organization are the most visible. Most organizations have written versions of rules, responsibilities, procedures, missions, and goals as well as a written budget. The people who work for and are served by the agency may or may not have seen these documents, but you should have little trouble looking at them if you ask.

Rules

Every organization has at least some formal written rules that everyone is supposed to follow. These rules can be found in an employee handbook or policy manual, which you may already have seen. Some agencies have very thick manuals; some have more than one. You may need to ask for help in finding the most relevant material. It is also important to know how these rules and procedures were established and by whom (Caine, 1998). Sometimes it is done by a committee; other times it is delegated to one person or written by a top administrator. The change process is also significant. Some policies have not been changed, or even reviewed, in many years. Other organizations have a regular method to review and update their policies and procedures.

It is also interesting to see how many people at the agency are really familiar with the rules. Both of our colleges have a student handbook; chances are that yours does, too. In our experience, students read that book when they have a question or when they think they may have broken a rule or been treated unfairly. Otherwise, they may never open it! At some placements, you will find that this is true for the policy manual as well.

Roles

Understanding the roles in an organization clarifies the division of labor. Roles describe the positions in the organization and the duties or functions that each one performs on a regular basis. The formal roles in an organization can be found in two places. Most organizations have job descriptions for each position that state, at least in general terms, the responsibilities of the positions. These descriptions may be found in the policy manual. They are often generated, or reviewed, when someone is hired. A prospective employee will usually want to see it, and in a large organization, a higher level administrator needs to see a job description to approve the hiring. Many placement sites even have written job descriptions for interns.

An organizational chart is also valuable when examining formal roles. An organizational chart does not describe the duties of each position, but it shows their position relative to one another. It shows who is responsible for whom and to whom each person is accountable. A chart for a hypothetical human services organization appears in Figure 7.1.

Organizational charts give a graphic representation of both horizontal and vertical differentiation in a particular organization (Queralt, 1996). Horizontal differentiation refers to the division of responsibilities shown on the chart. Most organizations have different departments, each with different responsibilities. In large organizations, there could be many people in each department; in very small organizations, some departments contain only one person. For example, at the Beacon Youth Shelter,[1] you can see departments devoted to functions such as teaching, child care, social work, and finance.

[1]Beacon Youth Shelter is a fictional agency that is used as an example throughout this chapter. It is an amalgamation of several agencies with which we have worked.

FIGURE 7.1
Organizational Chart: Beacon Youth Shelter

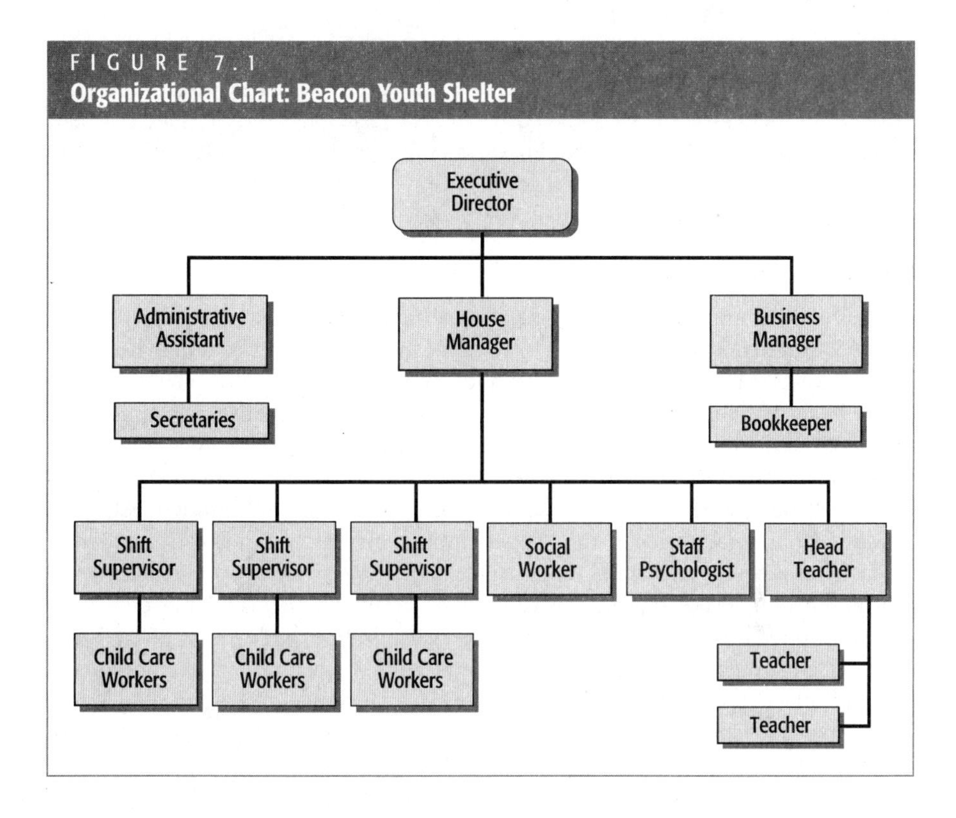

Vertical differentiation shows the chain of command. You can see from the chart that the top administrator is the executive director. In some organizations, the chief administrator is responsible to a board of directors. The executive director of Beacon Youth Shelter is responsible for all phases of the operation. However, most of the actual running of the shelter is done by other people. The executive director is most closely linked to the house manager, who is in turn responsible for several professional positions. The executive director may or may not attend staff meetings or know any of the child care workers by name. An organizational chart also gives you an easy view of the length and complexity of the chain of command. Organizations that have many layers and a long chain of command are referred to as *tall* organizations, while organizations with fewer layers and a shorter chain of command are referred to as *flat* organizations (Queralt, 1996). Each has its advantages and disadvantages.

You will notice that there is no intern on this chart. If an organization does not have an intern each semester or year, it may not include one. However, it is important to find out where you fit on the chart. Most interns tell us they fit "on the bottom." While that may or may not be true, it does not help us or the interns know whom they are responsible to or that person's place in the organization.

You will also notice that there is no place for clients on the chart. While that is not unusual, it is important to know the position occupied by clients in your agency (Queralt, 1996). In some agencies, clients are included in decision making. There may be an advisory board, for example, or a client council. This arrangement is more likely in organizations where clients can choose to use (and pay for) the service or go elsewhere. This arrangement gives clients some power and leverage. However, in agencies where the clients are not there voluntarily or have no other alternatives, clients may have very little influence on decisions.

Communication

Successful communication is, of course, the critical factor in making any organization run smoothly. In a small organization, communication may be very informal. Everyone sees and works with everyone else all day, and important information can be shared relatively easily, although it sometimes is not. Communication in larger organizations is more complicated. It may happen in the form of meetings, which can occur monthly, weekly, or daily. Some agencies begin each shift with a meeting. Various staff groups then meet once a week for more in-depth discussion, and the whole group meets once a month. Communication can also happen through memos, which is short for memoranda. Memos can be sent to just one person or to a whole group of people. In some organizations, employees receive several memos every day; some people call them "organizational junk mail." Increasingly, larger organizations with computer networks are using electronic mail for internal communication. Of course, lots of communication takes place through personal conversations, which can be planned or unplanned.

It is possible to look at patterns of communication in an agency. In some organizations, communication flows from the top levels down, with each level playing a role, and from the bottom levels up, with relatively little communication among people or departments at the same level. This arrangement is called a *chain pattern.* In a *wheel pattern,* on the other hand, one person or department is at the hub. Most communication flows directly from various departments to this hub and from the hub to the departments. Again, there is relatively little communication among the departments themselves. In an *all channel network,* everyone communicates with everyone else (Queralt, 1996). There are many other configurations, and it can take some time to determine what pattern a given agency actually uses.

Decision Making

Suppose you are working at Beacon Youth Shelter and you have an idea for a field trip. There is an interesting exhibit at the science museum, and you are successful in interesting some of the residents. But the exhibit is not going to be there much longer. Who must you ask about this trip and how long will it take to get approval? Your supervisor thinks it is a great idea, but you are surprised to learn that the executive director, who is out of town, must approve all such trips. You also have to obtain permission from each child's parent, guardian, or case worker, and teachers must approve the absence from school for a day. This story is a bit exaggerated (although in some cases not by much), but it illustrates the importance of understanding the decision-making mechanisms at your placement and how they affect your work. What projects or interventions do you have planned that may require multiple approvals?

Evaluation

In most organizations, there is some mechanism for evaluating or assessing performance. Sometimes a new employee is given an initial evaluation after a few weeks. Periodic evaluations may be done annually, semiannually, or even monthly. Evaluations are usually conducted by the person to whom you are accountable (see the organizational chart). There are many approaches to evaluation. Some methods allow the employee to set goals and then be evaluated on how successfully these goals have been reached. Other times the criteria and goals are set by the supervisor, and there are lots of approaches between these two extremes. Some organizations use self-evaluation or peer evaluation in addition to or in conjunction with the supervisor's evaluation. The format of the report can also vary from a simple checklist or set of numerical ratings to more complex narrative formats.

Purpose

It is easy to gain a general sense of the purpose of your placement; sometimes the name is all you need, as in Planned Parenthood or Child and Family Services. There is, however, a wide range in the specific objectives, general philosophies, and values among organizations, even among those with very similar titles.

GOALS

It is important to know your agency's overall goal. Some agencies are run to make a profit. Yes, they provide valuable services, but they expect to make money as well. Other agencies are nonprofit or not for profit. Their goal is to serve clients and break even financially.

It is likely that your placement site has a written statement of its goals. Here are some examples from agencies we have worked with:

> To provide quality and multicultural services to those whose lives have been affected by sexual assault.
>
> To provide education directed at the prevention of violence.
>
> To serve adults and youth who exhibit or are at risk of criminal or delinquent behavior, substance abuse, or mental illness, as well as other socially disadvantaged persons.

These formal goals statements may be well known to the people who work at the agency and/or to its clients, or they may be unknown to either group. A lot depends on who determined the goals and through what process (Caine, 1998). There will most likely be greater investment in and awareness of goals that were arrived at collaboratively, as opposed to those created by one or two people and then handed down.

HISTORY

Every organization has a history, and it provides an important context for the way things are currently being done. Internship placement sites range from organizations started by a grant from the federal or state government to long-standing charitable organizations and corporations. Older organizations have often gone through a number of changes. For example, the Beacon Youth Shelter always had a rule that they would not accept residents with a history of physical violence. However, they have recently agreed to change and allow children with violent histories to come in. That means this population is relatively new to them, and there is no doubt an adjustment period going on for everyone. Can you see why this information would be useful to an intern?

VALUES AND PHILOSOPHY

Each agency has a basic approach to working with clients. Consider, for example, a controversial program dealing with very violent children, whom many other agencies and in-school programs have been unable to help. The agency's basic philosophy is that most of the time these children use violence or the threat of violence to avoid taking responsibility for themselves. If students have not done their homework, for example, and are confronted by their teachers, they may erupt. If students are acting out in class and are punished, violent outbursts may result in everyone's attention being directed toward the violence, and the original infractions are forgotten.

The program also believes in giving children choices and holding them responsible. So, if students are acting out, they are asked—once—to stop. If they do not, they are not asked again; they are given a consequence, such as being forced to stand absolutely still, in a small room, facing the corner. They can say they are sorry or that they will stop now or threaten violence, but the consequence is still given. Furthermore, a threat or a violent episode only results in longer time in the room. Sometimes these situations escalate to the point where a child needs to be physically restrained. Once the child is calm, the original consequence is instituted. You may or may not agree with this approach, but perhaps you can see how these unusual methods flow logically from the program's philosophy and beliefs. These methods, in turn, are a lot easier to understand if you are aware of and understand the philosophy behind them.

Agencies also have values, or principles, that they try to adhere to in working with clients. Again, there are often formal written statements of these values, which may include such general principles as empowering clients, promoting self-sufficiency, or the right to privacy and confidentiality. Here are two examples:

> We use the word "survivor" rather than "victim" to describe someone who has been sexually abused or raped, in order to reflect the person's strength and healing capacity.

> The agency remains committed to extensive collaboration within the communities it serves. The systemic approach of the organization is best served by collaboration, given the complexity and nature of the issues confronted by our clients and in our communities.

Resources

The organization's budget may not seem very important to you; after all, you are probably not getting paid! However, the financial as well as human resources that an organization has determine what it can do for clients and often affect the general tone of the workplace.

Your placement site certainly has an annual budget. The budget may be part of a larger budget, as in the case of a senior services center that is funded by the city; its budget is part of the social services budget, which is in turn part of the city budget. The budget is usually broken down into categories, such as salaries, benefits, supplies, food, transportation, and so on. It is also important to know how the budget is set, who must approve it, and how changes are negotiated. Of course, budgets can change suddenly; in human services, that usually means they get cut, but occasionally an agency will be awarded a grant or get a new program approved and funded, which may in turn affect the agency's goals.

One aspect of the budget that is often overlooked by interns is where the money actually comes from. Even if your placement site does not charge for its services, they are not free. The money to pay the staff and administer the agency that provides the "free" services has to come from somewhere. Sometimes the clients are charged a fee on a set scale or one that varies according to their income. Sometimes there is a large nonprofit organization, such as the Salvation Army, that provides some of the funds. Organizations like the United Way also provide charitable funds to agencies. Other sources of funding include school systems or insurance companies, which may pay all or some of the costs of treatment. If the organization is public, like a school or an employment office, the funds come from taxes. Of course, many organizations depend on a combination of these funding sources.

This issue is important because the source of funding has great power over and influence on the organization. Insurance companies, for example, usually insist on a formal diagnosis before authorizing treatment. Some human service providers are troubled by these diagnostic labels, which tend to stay in clients' files and follow them around. Nevertheless, a diagnosis must be made, and it must be one the insurance company will pay for, or the client will have to pay cash or be referred to another agency. Organizations funded by tax dollars are vulnerable to the opinions (informed or not) of the taxpayers and must worry about public relations and influencing the local political process.

THE INFORMAL ORGANIZATION

As you read the section on the formal organization, you may have occasionally wondered how to answer some of the questions about your placement. In some cases, you don't know the answer at all, but in others, you may be wondering, "Do you want to know what's written down or do you want to know what's really going on?" Often, the formal rules, roles, goals, and values only tell part of the story. Consider the speed limit on the interstate. The formal limit may be 55 or 65 mph, but how fast you can go before risking a ticket is the informal rule that most people concern themselves with. If you live in a college dormitory, there is probably a time after which you are supposed to be quiet. But the unwritten rule usually is that you can make all the noise you want as long as no one complains. These are examples of informal rules that contradict or modify the formal rules. Other informal rules are not products of the formal rules. Most schoolchildren can tell you that it's wrong to "rat" or "tattle," often to the chagrin of adults. This is a very powerful rule for many children, yet it is not written anywhere. These are just a few examples.

In every organization, there is a whole network consisting of informal rules, roles, values, communication patterns, and all the other components of the formal organization. Both the formal and the informal aspects of organizations are important to understand, but the informal is often the most powerful and the hardest to see. In large organizations, such as state-level human service bureaucracies, the informal network provides people who work there with a sense of security and identity, because it is here that they learn their place in the organization and what is expected of them (Gordon & McBride, 1996).

Rules and Roles

Every organization has unwritten rules. Think back on jobs you have had and on your current placement. There is an official start time, but there is usually a grace period of at least a few minutes. In some places, you are expected to work late—for no extra pay—and in other places, you are not supposed to do that even if you want to. Many organizations have an informal dress code to go along with the formal one. Sometimes the informal rules come about because the formal rules have eroded or became obsolete, but have never been changed (Gordon & McBride, 1996).

At one family services agency, for example, the hours were 9 A.M. to 7:30 P.M., Monday through Friday. However, in recent years, the agency has been doing more and more outreach programming, working with groups in the community, meeting during the evening in church halls or people's homes, and developing a lot of weekend programs. Fridays are an especially busy day, as the finishing touches are put on the weekend activities. As a result, Mondays have become an unofficial day of "downtime." Few appointments or meetings are scheduled, and many staff members do not come in until noon. If you were an intern there and wanted to work on Monday or tried to launch a new group that meets on Mondays, you would receive a chilly reception, and you might wonder why.

Here are some other examples of unwritten rules that interns have encountered:

- Go to lunch with your co-workers or they will think you are a snob.
- Never go out to lunch; work at your desk.
- Don't miss the director's holiday open house.
- If you are dating another staff member, keep it quiet.

The written rules are usually easy to find, but how do you go about finding the unwritten ones? First of all, you will usually know when you have broken one. Just like in school, when you wore the "wrong" thing, used an expression that was suddenly "out," or sat at the "wrong" table in the cafeteria, people will react strangely; perhaps someone will even take you aside and clue you in when you violate an unwritten rule. Another way to discover the unwritten rules is to know the written rules and observe what happens when they are broken (Caine, 1998). In some cases, nothing happens. Either the rule has been bent in a way you don't know about, or it has been ignored altogether.

There is also a network of informal roles and relationships. It will not take you long to notice that some people perform duties that are not in their job descriptions. It may be that they were asked to do it and they didn't feel that they could say "no." Consider, for example, the worker whose supervisor discovers that her route home takes her past where his son has piano lessons. One day, the supervisor asks if she would mind picking up his son, since he will be in a meeting. The worker is happy to help, but it soon becomes an expectation.

Another way job descriptions expand is when something is not in anyone's job description; someone just decides to do it, and after a while everyone relies on this individual. There may be one person, for instance, who remembers everyone's birthday, takes up a collection for a gift when someone gets married or retires, or makes the arrangements for parties. Other roles in organizations might include the jester, who makes everyone laugh, and the go-between, who seems to have access to just about everyone.

Sometimes people expand their job description to make their jobs more interesting, especially if they have had the same job for a while (Gordon & McBride, 1996). At the Beacon Youth Shelter, a child care worker became interested in grant writing, volunteered to work on some grants, and eventually became quite good at it. Of course, his job description expanded and his paycheck did not, but it did make him a more influential person in the organization, which is another reason people sometimes try to expand their roles.

Both the formal and informal roles are connected by informal networks of influence. Secretaries often appear at the lower levels of the chart; in theory, they have little power and influence. However, it is often secretaries who know where everything is and how to navigate the paperwork maze. They overhear a lot of conversations, and they often control important resources, such as supplies. Often, if you want to know how to get something done, you are better off consulting an experienced secretary than the policies and procedures manual. In some instances, people on the staff are related (or were related) to each other. In other instances, they have shared decades of connection through schooling, neighborhoods, or churches.

Determining these informal influence networks takes time and reflection. You might try arranging everyone in order according to the influence they seem to have on decisions or on daily operations. Russo (1993) suggests that you interview people in the organization and ask them to whom they give advice, to whom they give orders, and from whom they accept advice and orders. This may be impossible in practice, and not necessarily well received by the staff, but if you could do it, you could then construct a web of connections that would tell you a lot about how power and influence flow in the organization.

Finally, there are informal subgroups or cliques in many organizations. At Beacon Youth Shelter, for example, there is a group of child care workers who have been there a long time; they are referred to as "the veterans." Few of them had any formal education to prepare them for their work; they learned on the job. Some of them rose in the organization to become shift supervisors, and until recently, the house manager position was held by a former member of this group. In recent years, though, there have been more and more workers hired who were graduates of programs in human services or social work. They are often younger and full of new ideas. Some of these ideas work, but others don't. The veterans sometimes shake their heads at these "whiz kids," and sometimes the whiz kids get impatient with the veterans.

Of course, both groups have a valuable perspective to offer the organization, and most of the time they work well together. Still there is some tension and competitiveness between them. It doesn't help that the new house manager did not come from within, is a graduate of a human services program, and has a master's degree. The veterans miss the easy relationship and open access they had with the old house manager; they could just approach her privately and accomplish a lot. Not only do the veterans not have this relationship with the new manager, but some of the whiz kids seem to be developing this informal access.

Communication and Decision Making

You can see from the foregoing description that sometimes communication and decision making do not follow formal lines and procedures. Informal avenues of access and influence can have a great impact on decisions. When this happens, other people in the organization are sometimes unaware of why the decision was made and/or are surprised by the decision itself. There are also blocks to communication. An organizational chart may show you who should be communicating with whom, but that doesn't always happen. Staff members who do not like or are resentful of their supervisors may in fact withhold important information. Other times a person in a formal role may be bypassed in favor of someone who performs the same function informally.

At Beacon Youth Shelter, each child is assigned a child care worker to be his or her "primary." One of the newer workers is not working out as well as had been hoped, and his assigned residents have been complaining. Some of them have been transferred to other staff. Of course, this happens from time to time and is not normally a problem. But with so many wanting to transfer, it creates a problem for other staff, who must handle more than their share. So for now, the transfers are being discouraged. However, the worker occasionally handles problems with his assigned residents in a way that causes an outburst, which is disruptive to everyone. So, other staff members have taken to counseling these children on an informal basis. If the child has had a problem in school, for example, it may not be entered in the shift log until after another staff member has helped the child resolve the difficulty. That way, by the time the primary worker knows about the problem, it has been solved. You can probably see the wisdom—and the dangers—of this approach, but it is just one example of the informal arrangements that mark all organizations and how they evolve.

Evaluation

The way in which staff members are evaluated says a lot about what is important to the organization. One important question is: What aspects of the job are included in the evaluation? Think about your classes in college. You have readings assigned, and many of your instructors may tell you that class participation is important. However, if you are never tested on your understanding of the reading and participation is not a part of your grade, you may be tempted to let these areas go, especially if they are difficult. By evaluating you on certain components of your performance but not on others, the instructor is sending a message, perhaps unintentionally, about what is important in the class.

Returning to your placement site, staff members are supposed to be evaluated based on how well they perform the duties in their job descriptions. In practice, however, some of these duties may not be reviewed, commented upon, or counted toward promotion, pay increases, and other organizational rewards. In addition, a supervisor may comment on aspects of the job that are not in the job description. One worker we know was criticized in writing for not being sociable with others at the agency, even though his cases were handled quite well.

How you are evaluated as an intern is probably established formally. There is either a form developed by the site or by your academic program. However, your supervisor's understanding of those forms can affect how you are evaluated. For example, a supervisor who believes that the purpose of your internship is to develop your helping skills may be more concerned with your grasp of theory and interventions and pay little attention to your efforts in understanding the larger context of the agency's work or your capacity for administrative roles.

Some organizations evaluate performance, some evaluate outcomes, and some evaluate both (Caine, 1998). Again, an analogy to school may help illustrate this distinction. Teachers of a U.S. history class may be evaluated based on whether they cover all the material by the end of the year and whether students seem to enjoy and be stimulated by the class. Another approach would be to evaluate teachers based on what students are learning. Some schools believe that it is the teachers' responsibility to lay out the information and the students' responsibility to learn. Others believe that the school and its teachers have a responsibility to see to it that every child learns. The evaluation methods chosen reflect these two approaches. Similarly, a human service agency may evaluate employees based on the number of clients seen or the number of hours worked, or it may choose to find ways to measure the change in clients and evaluate its workers in that manner.

Goals and Values

Earlier in this chapter, we discussed the formal statements of goals and values that most organizations have. However, in observing the organization, you may find that it has goals and values that are very different from the ones that are written down. The informal goals of an organization are also referred to as *operational goals* (Gordon & McBride, 1996). These goals are the standards by which day-to-day decisions are made, and they are sometimes not the same as the formal goals and mechanisms.

Beacon Youth Shelter has a stated goal of integrating clients back into the community. However, for some clients, the available community options are not good ones, even though they are technically available. The staff believes that the interests of these clients would not be served by such placements. So, the operative goal becomes keeping these clients in placement as long as possible. This goal can lead to a different set of conversations with funding agencies and case workers than the formal goal.

Operational goals can also be thought of as the "rewarded" goals. Look at how and for what the agency and its staff are rewarded. Are they paid by the client and therefore rewarded for seeing the maximum number of clients possible? Are they rewarded for staying within their budget? These rewards will affect the day-to-day operations of the organization.

It is also possible to examine operational values. These values are deduced from what is done rather than from what is written. Language is one key to operational values. For example, consider the way staff members talk about clients and the labels they apply to them. Sometimes staff members will refer to their clients in openly derogatory terms, like "retard," "dummy," or "geezer." Other times the labeling is more subtle. For example, sometimes certain clients are referred to as "difficult" or "hard cases." When we hear this, we are always tempted to ask: Difficult for whom? This comment is made using the experience of the staff as a base for description, not the experience of the client. The same client could be described as "having a hard time accepting authority" or "unwilling to accept personal responsibility." These terms are much more descriptive of the *client's* experience. Of course, people are going to become frustrated and say negative things from time to time, although preferably not to the client. A pattern, however, of one sort of comment or label can be a clue to an unconscious organizational value.

In the widely acclaimed television series *M*A*S*H*, the coping needs of the characters were such that facetiousness and depersonalizing became their lifeline to sanity. In some human service agencies where crisis casework is the majority of the load, the "*M*A*S*H* syndrome" becomes an accepted means of survival.

Another clue to organizational values is the way time is spent. We often say, "I don't have time," when what is true is that we choose to spend our time in a different way. All agencies and organizations have more demands on their time than they have time to meet them. Choices must be made and those choices reflect values. If a resident runs away or is injured, the agency may or may not choose to interrupt the daily routine to help the other residents deal with their feelings. Some agencies set aside time to meet regularly with staff for support purposes; others do not. When a decision has to be made, some agencies will consult with clients before acting. In some organizations, a good deal of time is spent planning for the future, while others reel from crisis to crisis and cannot seem to find the time for planning.

Most organizations put some time and energy into helping their staff grow as professionals and do their jobs better. These activities are often referred as *staff development.* Staff development may take place during a portion of regular staff meetings, or special meetings may be held exclusively for that purpose. Some agencies take their staff away on retreats. Others send them to workshops offered by colleges or training organizations. Still others support their staff in continuing their education. The connection to values is not just in how much time and money is put into staff development, but in the purpose of the activities (Caine, 1998). Some activities are directed at helping people learn to do their current jobs better. Others are directed at learning a new set of regulations, policies, or procedures. Still others are aimed at helping the staff learn more about theories so that they can come up with and share applications to the organization's work.

Finally, some staff development activities are directed at the career development of the staff. These activities are undertaken with the understanding that most of the staff will not and should not stay in their current jobs. They may move up in the agency or they may move on to a parallel, or better, position in another agency. Staff development of this kind is an investment in an individual and the profession as well as in the current functioning of the agency.

The informal workings of an organization are both fascinating and elusive. They are not easy to discern, but they are very powerful. At the end of this chapter, we have included several exercises for discovering the informal organization. Choose among them with great care and in consultation with your supervisor and instructor. Some of them require you to ask questions that may seem intrusive or to get information that may seem private. Remember, the idea is to enhance your internship, not to make it more difficult or complicated.

THE EXTERNAL ENVIRONMENT

You may already be beginning to see how the world outside the agency has an impact on its functioning. The number and nature of the organization's relationships with other organizations and agencies, sometimes called the *task environment* (Queralt, 1996), will affect your work and the ability of your placement site to do its job. In a more indirect but no less important way, the surrounding community, the economic climate, and some political issues affect you, your clients, and your agency. This network of influences is referred to as the *general environment* (Queralt, 1996).

The Task Environment

Most organizations cannot function without smooth relationships with other organizations. This is certainly true in human services. For example, the Beacon Youth Shelter works with two state agencies that refer and pay for residents. A local food bank helps supply their food. A large clinic in the area contracts with Beacon for certain psychological services. Some of the students attend school off grounds. Occasionally, a resident runs away or causes some problem in the community; good relationships with neighbors and the local police are essential. In addition, residents do not stay at Beacon forever; sooner or later, a new placement must be found for those who cannot return to their families or live on their own. This aftercare placement process involves detailed knowledge of and good relationships with a variety of other placement sites.

Furthermore, most human service organizations have one or more agencies that monitor them in some way. Some agencies are part of a larger parent organization. That is not the case with Beacon Youth Shelter, but other shelters in the city are funded by organizations like Catholic Charities or the Salvation Army. If your agency receives money from the state or federal government, chances are that it is also reviewed periodically by a government agency. Beacon Youth Shelter, for example, is a residential placement that contracts with the state social service agency. Hence, there are teams of people from the State Office for Children and the Department of Social Services that come and review or inspect Beacon. In most cases, a lengthy set of documents must be prepared for these visits.

The General Environment

There are a number of political issues that affect placement sites; we will just mention a few of them that seem to be common across many different placements. Local politics can have a considerable impact. For example, in some states, there is a cap on property taxes, which make up a large portion of each town's budget. Towns can vote to exceed the limit, and these votes are often taken to provide a special service, such as a new gymnasium or road repairs. If your placement is heavily funded by the town, these are important issues for your agency.

The attitudes of the people in the town toward your agency, its clients, and the work you do are important. Some agencies, like group homes, seem to inspire at least apprehension, if not fear, among neighbors. People agree on the need for them, but want them located somewhere else. If your agency has just opened up in a hostile or unwelcoming community, it has a large public relations task ahead of it. Some agencies make deliberate efforts to involve people in the community in their work. This is especially important when the clients are residents of that community, as they are in a Planned Parenthood Center, for example. If these residents are suspicious of and hostile toward the agency, chances are that fewer clients will come for services.

State and federal politics can have an impact as well. A new governor or mayor often means new people in charge of government agencies, and sometimes means a change in philosophy, along with new regulations and changes in the way funds are distributed. Federal issues such as welfare reform and abortion rights may also have a large impact on your clients and the mission and goals of your placement site.

SUMMARY

In this chapter, we have tried to introduce you to some ways of thinking about organizations. Although you may feel somewhat overwhelmed by the information, we have really only scratched the surface. Systems theory, organizational dynamics, organizational development, values, and staff development are enormous topics. In our experience, many interns find these topics fascinating. Others do not find them so interesting. However, we believe that the concepts we have introduced here, at a minimum, will help you understand both your placement site and what is happening to you there. At the very least, this chapter should stimulate some thought now and be a resource if you begin to have problems at the placement.

For Further Reflection

These activities ask you to gather some basic information about your agency and to think about what you have observed there. You will undoubtedly have to ask your supervisor or other agency personnel in the case of some basic information. If the person you ask doesn't know, ask where you can find out.

1. Where are the formal rules of the agency located? Have you looked at this resource? Have you observed any informal rules? Do any of them conflict with the formal rules?
2. Obtain or create an organizational chart of the agency. Are interns on the chart? If not, place them in an appropriate location.
3. Have you observed instances of people performing tasks and functions that are not in their job description? What do you know about how this happened? How about informal roles such as "office clown," which aren't in anyone's job description? Do there seem to be informal or unwritten expectations of you as an intern? How do you feel about those expectations?
4. How are decisions supposed to be made at your site? What evidence do you have that decisions are or are not actually made that way?
5. What is the agency's overall goal and philosophy concerning the clients? Working with clients? Managing employees and interns?
6. Have you been able to observe any operational values as manifest in language, use of time, or some other internal dynamic? How do they compare with the stated goals and values?
7. How does your agency handle staff development? What values are expressed in their approach?
8. How many years has your agency existed? Have there been any major changes during that time? Describe them.
9. What is your agency's annual budget? Where does the money come from? Do clients have to pay? If not, does some other agency pay? If the services are "free," where does the agency get the money for payroll and operating expenses?
10. What are some of the agencies with which your site has important relationships? What do they do?
11. Are there any agencies that monitor or control your agency? Are there periodic site visits?
12. What have you been able to learn about the relationship between your agency and the surrounding community?
13. What local, state, or national political issues are affecting your agency and its clients?

For Further Exploration

Berger, R. L., & Federico, R. C. (1985). *Human behavior: A perspective for the helping professions* (2nd ed.). New York: Longman.

A thorough discussion of systems theory, including many concepts not discussed in this chapter.

Egan, G. (1984). People in systems: A comprehensive model for psychosocial education and training. In D. Larson (Ed.), *Teaching psychological skills: Models for giving psychology away* (pp. 21–43). Pacific Grove, CA: Brooks/Cole.

Egan, G., & Cowan, M. A. (1979). *People in systems: A comprehensive model for psychosocial education and training.* Pacific Grove, CA: Brooks/Cole.

A wonderful book that integrates a systems and individual development approach to human services. Excellent chapters on systems. Unfortunately, the book is out of print. If your library does not have a copy, try the preceding Egan reference.

Gordon, G. R., & McBride, R. B. (1996). *Criminal justice internships: Theory into practice.* Cincinnati, OH: Anderson Publishing.

An excellent chapter on organizations from a criminal justice perspective. Covers both public and private settings.

Queralt, M. (1996). *The social environment and human behavior: A diversity perspective.* Needham Heights, MA: Allyn & Bacon.

A social work text with an excellent chapter on organizations.

Russo, J. R. (1993). *Serving and surviving as a human service worker* (2nd ed.). Prospect Heights, IL: Waveland Press.

Interesting chapter on organizations, with examples from criminal justice and mental health settings.

Senge, P. M. (1990). *The fifth discipline: The art and practice of the learning organization.* New York: Doubleday/Currency.

Very interesting and accessible discussion of systems theory and organizations. The applications are business oriented, but the concepts are very clearly explained.

References

Berger, R. L., & Federico, R. C. (1985). *Human behavior: A perspective for the helping professions* (2nd ed.). New York: Longman.

Caine, B. T. (1998). Using Bolman and Deal's four frames as diagnostic tools: Key concepts and sample questions. In B. T. Caine (Ed.), *Understanding organizations: A ClassPak workbook.* Nashville, TN: Vanderbilt University Copy Center.

Egan, G., & Cowan, M. A. (1979). *People in systems: A comprehensive model for psychosocial education and training.* Pacific Grove, CA: Brooks/Cole.

Gordon, G. R., & McBride, R. B. (1996). *Criminal justice internships: Theory into practice.* Cincinnati, OH: Anderson Publishing.

Queralt, M. (1996). *The social environment and human behavior: A diversity perspective.* Needham Heights, MA: Allyn & Bacon.

Russo, J. R. (1993). *Serving and surviving as a human service worker* (2nd ed.). Prospect Heights, IL: Waveland Press.

SECTION THREE

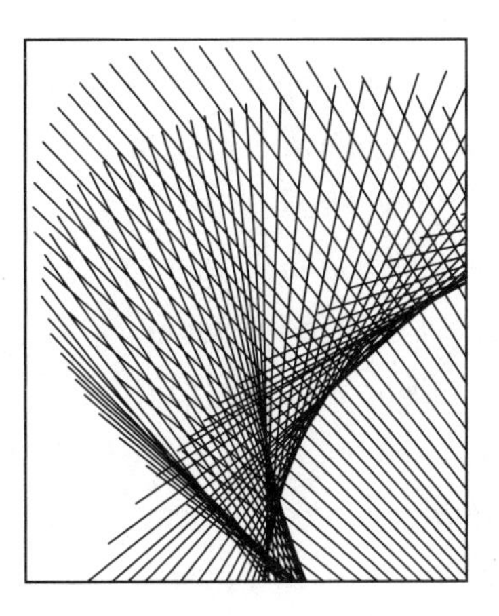

MAKING IT BETTER

In the preceding section of the book, we helped you prepare for, examine, and work through many different issues that arise in the beginning stage of an internship. Once you are finished with those issues—or at least once they are not foremost in your mind—you can turn your attention to other things. This section of the book is designed to help you keep your progress moving and your internship alive and vital.

In Chapter 8, we look at some common traps that can await you at this point in your internship. We also encourage you to take a careful and thoughtful inventory of the progress you have made and the issues that remain for you to resolve. And we explore a common, although not universal, phenomenon: the feeling that your internship is falling apart. Finally, in Chapter 9, we present you with a model for thinking about and dealing with the obstacles and difficulties that you have identified.

This section is less theoretical than the previous one. Our primary objective is to get you to examine your own experience, and we believe you already have the theoretical tools with which to conduct such an examination.

Although you may find the reading in this section a bit easier, the issues you will deal with are every bit as difficult as those faced in the Anticipation stage. However, if you have worked through the Anticipation stage, then you know the sense of satisfaction and empowerment that comes with meeting issues head on. In the previous section, you achieved an informed, realistic commitment to your work. In this section, you can achieve a clearer vision of that work and a greatly increased sense of confidence that, through reflection, effort, and reaching out for help and support, you can handle whatever comes.

CHAPTER 8

Considering the Challenges

I knew I would learn an incredible amount because of the actual hands-on experience that you cannot get from a book. But I was surprised at the impact the internship had on me and to learn how much I needed to work on.

STUDENT JOURNAL ENTRY

You have dealt with the issues and concerns that accompany the first weeks of the internship and moved successfully through the anticipation stage. Now what? Do you relax and coast through to the end? Well, you might be able to do that (although it's not likely), but you will miss many learning opportunities if you do. Do you wait, apprehensively, for the problems to come? After all, you read about the stages of an internship, and disillusionment is the next stage. In our experience, almost all interns do experience some problems, but how many, how severe, and when they will occur are difficult to predict. Dealing with them is part, but not all, of the challenge of the next portion of your internship.

One focus of this book is to help you become a proactive learner, shaping your own learning experience as much as possible. So, what we want to talk about now is how you can keep yourself growing and moving forward. There are three major components to this effort: (a) taking inventory of where you are now and where you want to be, (b) identifying any problem areas that need attention, and (c) developing and implementing plans to take action in both of these areas. These steps of assessment, problem identification, and problem solving involve skills that are essential to the success of your internship, and you will continue to need and use them throughout your placement. Each of them, though, can involve and evoke powerful feelings.

In this chapter, we begin with some thoughts about growing and learning. We then present some ways for you to take stock of your progress so far. We pose some general questions about your internship and then help you think about issues and challenges in the major arenas of your internship: the work, the people, the site, and yourself. Then we invite you to think about the problems you may have had or anticipate having and how you feel about them. It is at this time in the internship when many people enter the Disillusionment stage, and we will help you make sense of that experience. Toward the end of the chapter, we help you think about what steps you may want to take in moving through the new challenges in your internship.

THINKING ABOUT GROWTH

The Human Side of the Internship

Why did you decide to do an internship? For many interns, the answer is the same as the reason they are going into human service: to help people. As time goes on, though, many interns become absorbed in the assessment techniques and intervention skills they are learning. Other internships are focused less directly on people in need, and in those cases, the overall goal may have been skill and career development right from the start. It may seem to you, then, that the overall goal of your internship is to inquire into the nature of phenomena—be they people, systems, or projects—by studying them firsthand in their natural environments and providing necessary services to them; for example, you might be providing services to elders in the community or to adolescents in crises with their families, their schools, or the law. And of course, you want to do the best possible job you can in providing such services.

However, we remind you that your internship is fundamentally a *human* experience (Georges & Jones, 1980). When the focus of internship is on collecting data, analyzing information, providing services, or even acquiring skills, rather than on *living the experience* of the internship, the effects of the human side of the experience on you and your work are easily ignored or only mentioned in passing, as if they were of little importance to the real work of the internship. Ironically, it is precisely this human element that makes the field experience one of mutual learning and vulnerability for you and for the others who are part of your experience in the field (Georges & Jones, 1980; Wagner, 1981).

So, how do you make sure you are still attending to the human side of the internship? First of all, you need to remain open to and accepting of your feelings, even those messy ones that you wish you didn't have or think you're not supposed to have. The more you are open to your own emotional experiences, the more you will be open to the emotional experiences of others. Your feelings are also valuable cues to learning opportunities. Second, you need to remain open to the relationships you have in your internship and the joys and challenges they bring. Finally, you need to remain open to individuals. Remaining open means listening, reflecting, paying attention, and resisting the temptation to put people—including yourself—in categories and boxes that limit your experience of them.

Experiencing Change

Moving on and changing are generally accepted as parts of the internship process. You expect them and so do your site supervisor and your instructor, so much so that if they do not occur, it can be cause for consternation. Moving on suggests taking on additional responsibilities that further test and develop your skills and competencies. You expect your workload to increase and your performance to improve. Essentially, you seek out greater challenges and strive to meet them, often with considerable satisfaction.

Changing, on the other hand, suggests that an internship affects you in noticeable and pervasive ways. During the first seminar class, just after field placement begins, we often ask our interns this question: What would you say if I told you that the person you are today, sitting here in class, is probably not the person you will be when you complete your internship? Our interns tend to look at us very strangely and quietly when we pose this prediction to them. How about you? What is your reaction to such a prediction? Would your reaction be different if the question were asked at the very beginning of your placement? Maybe the question strikes you as a little dramatic, and perhaps it is, but change involves something more substantial, more far-reaching, and more challenging than just taking on new responsibilities or improving your skills. It involves looking at yourself, your clients, the helping process, the placement site, other organizations, and even society in new ways, and this is not solely an intellectual process.

Change of this sort can be enormously exciting, and it can also be frightening. There are natural tensions involved in such change, and you are bound to feel them. According to Jon Wagner (1981), you deal with two sets of perspectives during your internship: the perspectives you bring to the field and the ones you are developing as a result of your work. On the one hand, you are drawn to the reality of the work, its vitality, its unpredictability, and its dynamic yet problematic character; on the other hand, you seek the comfort of "home," where you know the situations and the problems, and where you are known to others. Your challenge is to bring these two sets of perspectives together in such a way that learning goes on. If you keep both your feet safely planted at home—concentrating only on how you currently think and what you already know—you underextend yourself in the experience and risk not growing; however, if you bury yourself in unfamiliar soil, you overextend yourself and risk sinking in the experience.

Resolving these differences requires that you embrace both perspectives at the same time, "straddling them" in Wagner's (1981) words, so as to create a perspective or vision that incorporates the past as well as the emerging ways of looking at things. Sounds great, right? It can be, but William Perry (1970) reminds us of two opposing human urges: the urge to progress toward maturity and the countering urge to conserve. These competing emotions accompany the two perspectives just discussed and can interfere with their integration. Similarly, Robert Kegan (1982) believes that all change involves letting go of old familiar ways, and that can be frightening. So, expect some excitement as well as some trepidation, and remember that both these emotions can help you grow.

Given these challenges and emotions, how can you keep yourself moving forward? For one thing, you can remember the experiential learning cycle described in Chapters 1 and 2. Mobilize your own resources and those around you so you can seek and have new experiences. Remember, though, that action and experience are only part of the picture. The need to *do* is very high for most interns, as you have probably discovered, and the need of placement sites can also be high due to the sheer volume of work and reduced staff size. However, the activities of processing and reflecting are equally critical to your professional development and learning, and they need to be structured parts of your internship experience in activities such as journal time and discussion groups with peers as well as designated personal time for thought and reflection. When the balance between action and reflection is compromised, so is the overall quality of your field experience. How are you doing so far in your internship at achieving this balance?

You also need a balance of challenge and support from yourself and those around you. You need support for who you are now and the way you think and feel, but you also need to be challenged to move forward and take risks (Kegan, 1982). You need support as you experience the emotional side of the internship as well as challenge, so you can use those emotions to learn and grow.

TAKING STOCK OF YOUR PROGRESS

With those thoughts about learning and growth in mind, we turn to the process of assessing your progress. This process involves considering some general questions as well as some specific issues. One theme you will notice throughout this section is the reaction you have when something you have read or heard about appears in front of you, or within you, in real human terms.

Keeping the Contract Alive

Remember your learning contract? When was the last time you looked at it or thought about it? In Chapter 5, we described the learning contract as a living, changing document, but we also know how easy it is to set aside and forget after it is written and handed in. One reason for this neglect is that after a few weeks there are often some discrepancies between what you planned and what you are doing. You may find that some of your goals have been met, and hence the activities associated with those goals are no longer as interesting as they once were (although you may still be doing them). There may also be learning activities that you have not been able to do and may not get the chance to do after all. Agency personnel, needs, and priorities sometimes change, so that an activity that once seemed perfectly possible must now be foregone or drastically modified.

It is also quite possible that you have had some new experiences and some opportunities to become involved in areas that you were unaware of when you wrote your learning contract. These unexpected turns in the road can lead to exciting new directions, questions, and goals. If your learning contract is to serve as a valuable guide and point of mutual clarity for you, your site supervisor, and your instructor, you all need to work to keep it from becoming obsolete.

Expanding Your Knowledge

Some of your goals were knowledge goals, and you want to be sure you are challenging yourself, and being challenged, to acquire more concepts and think at new levels. The exact nature of these knowledge challenges is, of course, an individual matter, but there are some issues you may want to think about.

The internship is a tremendous opportunity to integrate theory and practice. You have probably studied lots of theories in your classes, and even given some thought to their application. However, as you leave the role of observer and become a more active participant, your theoretical knowledge changes as the result of trying to apply it (Benner, 1984; Garvey & Vorsteg, 1992). This process can be troubling; you may find that your theories don't seem to work. Could all of your professors and textbooks have been wrong? The integration of theory and practice is a normal challenge in human service work. Not all theories work equally well in all situations. Furthermore, it may take you a while to recognize your theories in action because they look different when acted out by real human beings with real problems. How has your experience so far expanded and challenged your theoretical base? If your old frameworks are not working, perhaps you need to request some readings to help you find more suitable ones.

A related issue is the formulation of general principles to guide you in your work. These principles are derived from your experiences in class, in the workplace, and in other field and life experiences (Benner, 1984). Often, interns can tell us about what they did in a certain situation and why, but they cannot answer general questions such as: What are you thinking about as you approach a youth in crisis? or How do you know when to press for a response and when to let a client be silent? They can, perhaps, quote textbook answers, but they do not feel a *personal* connection to those answers because they are not coming from their experiences. As you accumulate and process experiences, you can try to formulate some general ideas about what works for you, what works in an agency like yours, and what your strengths and weaknesses are in a given situation.

Expanding Your Skills

When we discussed learning contracts, we mentioned that there are three levels of learning. You are moving into the second phase, in which the focus shifts from learning about the work to learning how to do the work. So naturally, you will become more focused on your skills. When you wrote your contract, you probably wrote about skills in fairly general terms. As you begin to acquire new skills, though, you can think about skill development in more specific terms and specify smaller steps for yourself.

Dreyfus and Dreyfus (1980) found that students go through five levels of skills acquisition: novice, advanced beginner, competent, proficient, and expert. In addition, they found that these levels of proficiency reflect changes in three aspects of skills performance: (a) a shift from reliance on abstract principles to past experiences, (b) a change in perceptions from seeing a series of parts to seeing things as wholes with relevant parts, and (c) movement from being a detached observer to being an involved learner.

Their model also has been used to study how health-care professionals acquire and develop the skills to carry out their work effectively (Benner, 1984). Benner's findings are particularly useful when thinking about this new phase of your internship. As a novice, you had little understanding of the situations in which the work must be carried out. The basic skills that you brought to the field were generic and in need of *contextual meaning*, that is, knowledge of how to use your skills effectively in a particular situation. For example, knowing the names of the resources in a community and being familiar with the indicators of potential suicide do not necessarily mean that you know how to interpret a client's behavior and prioritize your responses to deal with a life-threatening crisis. However, by the end of your internship, you might be expected to develop the skills to respond effectively to just such a crisis. Certainly, it would have been helpful to have taken a course to prepare you for understanding the contextual meaning of your work, but that may be neither practical nor available. At this point, you may be part way down the road to contextual meaning.

COMMON ISSUES FOR INTERNS

Just as there were issues that most interns face in the Anticipation stage of the internship, there are new issues that emerge as that stage begins to fade. Some interns report that they encountered all of the issues, a few report encountering none, and most fall somewhere in-between. The issues arise in the same categories we used earlier in the book: the work, the people, and the site.

The Work

Work issues are those that center on the demands being placed on you, the responsibilities you are undertaking, and the opportunities you are being given. First is the issue of assignment of responsibilities, in which you seek a balance between over- and underchallenge in either the volume, depth, or scope of what you are asked and/or volunteer to do. Too much responsibility can be overwhelming and leave you feeling ill equipped to do the work; too few demands result in being underchallenged, which can be humiliating, frustrating, and disappointing (Burnham, 1981; Royse, Dhooper, & Rompf, 1996).

You may find that you are doing tasks that you weren't expecting and that are not in your contract. Sometimes exciting opportunities arise. The agency receives a grant, your supervisor learns of a new support group that is starting, or you fill in for someone in an emergency and discover you like what you are doing. There is nothing wrong with these changes as long as you keep your goals in mind. If those goals are changing as well, then a shift in activities makes sense. Do not be lured by the novelty of an activity or responsibility, though, that does not really fit in with where you want to go and what you most need to learn. Sometimes you get these different responsibilities because the agency needs help and you are there. Of course you will want to help, and you will be nervous about saying "no" to your supervisor. However, if what you are being asked to do is out of line with your basic goals, something needs to change. This is a time when you may need the support and assistance of your instructor.

If you are doing well, you may also be asked to do additional work beyond the time negotiated in your contract. Be flattered, but beware! Remember your other commitments and responsibilities, and don't take on too much.

Balancing your needs as a learner with the agency's needs for a worker can become even more complicated when you are employed, or offered employment, while interning. Again, this is extremely flattering, and the money is nice, but you are there primarily to learn. Working involves doing; learning involves doing plus time for reflection. The compensation for work is usually money; as a learner, you need time and energy from your supervisor. Potential problems can be managed if you clarify your needs as a learner and as a worker by discussing them with your supervisor and keeping supervision separate for the two roles.

The People

For most interns, the internship unfolds in the context of a web of relationships. In previous chapters, we helped you think about some predictable challenges of the early phase of these relationships. The relationships are continuing to develop, though, and new issues will undoubtedly present themselves.

ISSUES WITH CLIENTS

If you are engaged in direct work with clients, a great deal of your psychological, intellectual, and emotional energy will go into this aspect of your internship. By now you have probably moved past the basic concerns of acceptance, although if you or your agency receives new clients, you will have to navigate these challenges again. Now, as you and your clients get to know each other better, you will have new concerns about the nature of the population and your evolving relationships with clients.

Nature of the Population As you move from an intellectual way of understanding clients to emotional involvement with them, you are bound to have reactions to them and their situations. The challenge here is to understand your reactions and learn from them rather than letting them interfere with your work and your general well-being.

As you get to know your clients, their individual and collective life situations can be overwhelming. If you are a member of many dominant,[1] relatively privileged groups in society and have been fortunate enough to live your life thus far without major trauma, exposure to people whose lives have been and continue to be very difficult can be upsetting on some pretty deep levels. Some interns are coming face to face with this reality in human rather than abstract terms for the first time. As one intern said, "I'm overwhelmed by the sad stories of their lives; I can't get them out of my head. This is too much for me."

Often, interns who are the most touched and outraged by their clients' situations are those working with what are called *socially vulnerable clients* (Brill, 1998). These are people who live in poverty, people with disabilities, elders, children, and members of other groups that have been targets of discrimination and harassment. Even if you are not working with socially vulnerable clients, you may have had your first close encounter with issues such as abuse, disease, or even death, and that is never easy (Wilson, 1981).

As you come to know your clients and their situations better, the issue of empathy can surface again. In Chapter 4, we discussed the problem that some interns have finding common ground and establishing empathy with clients. If you are successful, though, you face a new set of problems. Understanding others' viewpoints about and

---

[1]Remember that we are using the term *dominant* to refer to groups with relatively more access to power and influence. This concept was thoroughly discussed in Chapter 2.

experiences in the world challenges your own, and such challenges can be threatening at times. For example, one intern from a middle-class background was working with juvenile offenders. She could not understand how anyone could commit crimes and think the behavior was right, although she understood that people make mistakes or act impulsively. As she listened to clients, though, and came to understand how hopeless many of them felt and how angry they were at a system that never seemed to benefit them, she began to understand their contempt for the legal system and the reasons for their own code of ethics. She was quite rattled to discover that something she thought she knew for sure was not as certain anymore.

Another set of issues arises as you become more aware of the behaviors your clients are capable of and perhaps witness them firsthand. Destructive, irrational, or violent behavior directed at peers, family, or loved ones can be difficult to hear about or see. Such behaviors can also be directed inward, and you may have seen clients engage in behavior that is physically or emotionally self-destructive. Again, you know this already, but you may be *seeing* it for the first time.

Finally, such behavior can be directed at you in verbal or physical outbursts that can be unsettling even for experienced workers. If you are interning in the criminal justice system or in a psychiatric institution, there is a heightened potential for experiencing physical violence in the field. Perhaps you have felt this danger, regardless of whether you have actually been the target of violent behavior. Perhaps, too, you were in a potentially dangerous situation without realizing it. For example, if you accompanied your supervisor in transporting a youth whose custody is with the state and who has a history of assaultive behavior, then you were in fact in a potentially dangerous situation. Perhaps the youth was very likable and did not seem dangerous. But later, the supervisor cautioned you that you sat too close or did something else that exposed you to danger. Once it is pointed out, it makes perfect sense, but the level of awareness required in these situations may come as a shock.

Your emotional reactions become problematic when they interfere with your learning or when you are unable to carry out your field responsibilities without much personal difficulty. Even experienced workers "take their work home" sometimes, but if you find that you cannot get the clients out of your mind and cannot concentrate on other things, a problem exists. That sort of preoccupation can also affect your behavior in the seminar class. We have noticed that interns who worked with violence and other forms of extreme emotional content tend to be more needy in class for airtime and connections with others (King & Uzan, 1990). They also tend to be more cynical in their perspectives and sometimes discount assignments and being in class, noting that they were needed in the field, doing truly important work. In addition, they often have a heightened awareness of their limits and those of the systems within which they work and can become preoccupied with loss, pain, and violence. If you find these behaviors emerging in class, even if you are not working in a setting that seems dangerous or emotionally charged, you may be struggling with your emotional reactions to your clients.

No two interns react exactly the same way. Some issues affect most interns, and most issues affect some, but no one can tell you where your struggles will be. Pay attention to your feelings and spend time with your journal. Bounce your ideas and reflections off of your supervisor, your instructor, and your peers. These are lessons that no one can teach you except yourself.

Your Relationship With Clients Time and interaction bring new challenges with clients. The specific issues will vary with the intern and the client population, and we cannot cover them all. Your instructor, supervisor, and co-workers are valuable resources for anticipating and dealing with the issues that arise in working with your particular clients. However, some challenges arise with great frequency, including issues of progress, boundaries, difficult clients, and dealing with diversity.

If you have developed a reasonably comfortable relationship with your clients, you may be wondering how and when you can move beyond what many interns call "chitchat" into some real issues. This concern is not limited to those who work one-to-one with clients. Interns at sites such as shelters, soup kitchens, or drop-in centers often report that clients are now accepting of them. They have stopped giving them funny looks or ignoring them and will sit and talk with them. However, as one intern put it, "When I try to talk about the jam he got into the night before, a difficulty at home, or his drinking problem, he changes the subject. If I keep it up, he shuts down, or just gets up and leaves."

As you spend more time with and around your clients, you may find them engaging in all sorts of behaviors that seem unproductive or even counterproductive. For example, they may refuse to talk; they may talk on and on about their problems but show no interest in solving them; they may agree to try new behavior and then not follow through. Upon examination, there are many ways to understand such actions, and it may or may not turn out to be unproductive. Initially, however, interns who are not making the progress they think they should with clients are frustrated and concerned. What sorts of issues lie beyond acceptance with the clients at your agency? To what extent are you able to progress toward those issues and to what extent do you feel stuck?

Another natural issue that arises with clients concerns setting and maintaining boundaries. When you set boundaries, you are making your clients aware of the limits of your relationship and what is and is not permissible in that relationship. Boundaries are very important in the development of a relationship, although clients may not always appreciate them. Many clients inhabit an interpersonal world in which they never know what to expect from people or in which they are dominated and exploited in their relationships. Boundaries help the client understand your role and what to expect from you as a person carrying out that role. They also let the client know where in your relationship there is flexibility. Ultimately, they help the client feel safe.

Clients will often test those boundaries, and that can be hard. Clients may ask you for personal information that seems both irrelevant and inappropriate. They may ask whether they can call you at home or ask you to stay with them beyond your shift or the time limit of their appointment. They may invite you to dinner or to visit their home or to attend a family wedding (perhaps even their own!). You can probably think of many more examples and may even have experienced some in your placement. Perhaps you have handled these situations with relative ease. But some interns report feeling stuck in these situations—they literally don't know what to say. Still others tell us that they knew these issues might arise, they knew how they should handle them, but when the time came, they had trouble actually setting the limit.

Some boundaries are very clear and unquestionable, such as sexual relationships with clients. If a client is attracted to you, even if the attraction is mutual, there must be no sexual relationship. Finding a sensitive but clear way to set this limit can be difficult, however. Other boundaries are more flexible and situationally determined. The issue of socializing with clients, for example, can be a complex one, especially if your services are delivered in the community (as opposed to in an agency) or your clients are from a culture that expects some social contact with helpers. There is a good deal of literature on the issue of boundaries, and experts do not always agree; we suggest some literature at the end of this chapter if you are interested. Agencies, too, have different norms and policies, and you should certainly be aware of the ones at your placement. Dealing with boundary issues is an important frontier for many interns.

Some interns are surprised to discover that there are clients they just do not like. Others become frustrated with the behavior of their clients and often refer to these clients as "difficult." In Chapter 6, we suggested some client behaviors that you might find difficult. Research has been published on this issue. For example, two studies conducted in the 1980s found similar results in surveying therapists about their perceptions of stressful client behaviors (Deutsch, 1984; Faber, 1983). Although the order was different, both studies identified the same five most stressful client behaviors: suicidal statements; anger/aggression/hostility; apathy or lack of motivation; depression; and client's premature termination. This research was not done with interns, but they represent more examples of behaviors that might cause you to think of a client as difficult. Or perhaps these behaviors don't seem as challenging to you as some others.

As discussed in Chapter 7, we are not all that comfortable with the term *difficult client.* While it may not always be intended by those who use the term, to us it implies that the difficulty rests solely with the client. While it is true that certain behaviors, behavior patterns, and issues would be troubling to anyone, the depth and breadth of the emotional reaction vary greatly from person to person. Thus, again, the question is not which clients are difficult, but which clients *you* are finding to be difficult. Perhaps you have already learned something about the sorts of clients or behaviors that are especially troubling. If not, maybe you can think of a client with whom you have trouble. In that case, you might have thought about just what it is that makes the client difficult for you. As we discussed in reference to expanding your knowledge, with time you can move from identifying specific clients to some tentative, general statements about your reaction to clients.

Finally, coming face to face with differences can raise some new issues for interns. Perhaps you are working with clients whose race, ethnicity, social class, lifestyle, stage of identity development, or any other component of their cultural identity is new and very different from your own. You know how you want to feel about diversity, but it is important to be open, at least with yourself, about the feelings you are actually having. Even if you are struggling with these cultural issues, you probably expected them and talked about them in some of your classes. What you probably did not expect to deal with are the icons that will shatter as you learn about people who are different from you when you expected them to be similar. For example, there may be the shoplifting clergy, the alcoholic physician, and the sexually abusing attorney. We have met them in our work, and you will meet them, too.

ISSUES WITH SUPERVISORS

In this critical relationship as well, you are probably beyond the initial acceptance concerns. However, many interns find themselves facing new issues, which can be troubling in and of themselves but which can also threaten the foundation of trust and acceptance you have established.

Some interns are initially so impressed with their supervisors that they idolize them. Such interns talk about their supervisors as if they were all knowing and perfect in every way. They strive to be just like them. But then the bubble bursts. The supervisor makes a mistake with a client, snaps at the intern, misses an appointment, or exhibits one or more of any number of foibles that all of us fall prey to at one time or another. There can be a real letdown then, and the intern has to find new, more realistic ways to admire the supervisor.

By now, you probably also know a good deal more about your supervisor's style (you may want to review the discussion of supervisory style in Chapter 6). You probably also know more about yourself and which aspects of your supervisor's style suit you well and which do not. It may be, too, that your needs are different from the first weeks of the internship. You may need less directive supervision than you did or more emotional support. This can result in an improvement in the relationship if the supervisor is naturally suited to the needs you have now, a problem in the relationship if the supervisor is not, or the continuation of a good relationship if the supervisor is able to adapt her or his style to your evolving needs.

You may also have had disagreements with your supervisor. These disagreements do not have to be arguments; they may be differences of opinion about a client, a project, or a strategy. Acknowledging and discussing differences are considered good practices in the helping professions. However, doing so can create some anxiety for interns (Wilson, 1981). In most cases, open discussion strengthens the supervisory relationship as long as your supervisor responds empathically and educationally to your concerns. The question of how you and your supervisor will deal with differences is another style issue that you may want to discuss, if you have not already. Still, for most interns, the first time a disagreement occurs is at least moderately stressful.

As you experience the supervisory process, there may be issues that arise here as well. In Chapter 6, we asked you to think about how you were reacting to criticism. Well, you have had a lot more of it by now, and your reactions may have changed as well, for better or for worse. Another interesting issue for interns is how they respond to positive feedback. Some accept it gracefully; others seem uncomfortable, as if a bug has landed on their shoulders and they need to flick it away. Often, these interns find a way to deflect a compliment by giving someone else the credit, saying there was nothing to it, or bringing up a shortcoming. Do you seem to fall into one of these categories?

By this time you have probably had an evaluation as well. In Chapter 6, we encouraged you to learn as much as you could about the process so that some of the fear that comes from the unknown is alleviated. Even so, once you actually have an evaluation, you are going to have some sort of emotional reaction to it. This can be a time when you receive considerable affirmation about your performance as well as your potential as a professional. However, for many interns, this is also a period of introspection, self-doubt, and self-confrontation, which can raise painful issues. For example, your limitations and gaps in learning, skills, and style are identified openly during evaluations, and you may become preoccupied with your shortcomings, especially if you receive more negative feedback about your personal style than about your professional skills. Evaluations may cause you to scrutinize your skills, abilities, commitment to your field of study, and even to question the value of the work itself and your choice of profession (Lamb, Baker, Jennings, & Yarris, 1982).

Reading your evaluation is a little like getting back a paper. Some students read all the comments because they want to learn and grow regardless of their grade. Others look to see whether they got the grade they wanted. If they did, sometimes they don't even read the comments. If they did not, they immediately become defensive and upset. We hope you can take the former attitude. Even so, you may have a variety of reactions, whether positive or negative, including speechlessness, disbelief (some students cannot believe how good their evaluations are), anger, and tears. Some students who have gotten As in their classes may be shocked at being critically evaluated for the first time. It may be that they did very well, but the supervisor also noted areas for improvement. It may also be that they are better in the classroom than in the field, at least at this point.

Some interns who are doing a second internship and received high ratings previously are surprised if their current ratings are not so high. Perhaps the previous supervisor was unwilling to be critical, for any number of reasons (Wilson, 1981). If there are other interns at your placement, especially if they are from your campus, there may be some subtle, or not so subtle, competition concerning evaluations. If you think your supervisor was harder on you than a peer, or if your peer had another supervisor whom you think is an "easy grader," you may have trouble staying focused on learning from your own evaluation.

Whatever your reaction, we encourage you to take some time and find a supportive audience in and/or outside of seminar class to express your feelings. Especially if you are upset, blowing off some steam can help you clear your head and focus on what you want to do. A disappointing evaluation can also be seen as an opportunity. Even if the work did not come as easily as you had hoped, you can make a plan to further improve your abilities.

Many interns are reluctant to question an evaluation, but if you don't understand why you received a rating or comment, it will be very difficult to learn from it. One approach we find effective and comfortable is to have interns use a copy of the evaluation form to evaluate themselves. They can then compare their forms with their supervisor's as a tool in supervision. Keep in mind that you have a right to a high-quality evaluation, and if you do not receive one, you should consult with your instructor. High quality does not necessarily mean favorable; we have both seen very positive evaluations that we thought were poorly done. Suanna Wilson (1981) suggests that a high-quality evaluation is concise, specific, describes behavior (as opposed to using labels), and contains both positive comments and areas for improvement.

ISSUES WITH FACULTY INSTRUCTORS

In many ways, the same issues that arise with site supervisors can arise with the instructor who is overseeing your placement and/or conducting the seminar class. You do not spend anywhere near as much time with this person as you do with your site supervisor, but it is an important relationship nonetheless. This relationship can also have the added dimension of grading; in most programs, the instructor is partially or wholly responsible for issuing grades. Now is a good time to reflect on how that relationship is going.

ISSUES WITH CO-WORKERS

By now, you have probably forged a relationship of some sort with your co-workers. Some have accepted you and included you in their work and conversations. Others may pay little attention to you. Your relationships with individual co-workers will change over time, and it is hard for us to predict what issues will arise and with whom. However, there may be issues of influence and control that arise and manifest themselves in several ways. Some of your co-workers, for example, may try to exert influence over you by giving you orders and directions. This can be confusing, and you can end up feeling like you have more than one supervisor. If a co-worker is telling you something that is directly opposed to what your supervisor has said, you have an additional problem.

Co-workers may have hidden motives for befriending you, and it may take you some time to figure out what is going on. You might have to distance yourself from co-workers who are overly eager and overly accommodating until you are able to understand their motives. Often, these members of the staff are most marginal and most likely to reach out for alliances. In addition, if there are factions at the site, people may try to befriend you to recruit you to their side or clique.

Interns frequently tell us how important it is to feel like a part of the organization, a part of the staff team. It is an issue of acceptance early on, but develops into much more as the internship takes on its own life. Being part of the team is a vote of confidence by the staff that signals your status as an emerging professional. It happens when you create a community of mutual support, encouragement, work, and challenges that invigorates and sustains you in your field experience. On the other hand, interpersonal conflict with one or more co-workers or the general feeling that you are not a team member can be a major source of stress for interns (Yuen, 1990).

Your co-workers can also serve as role models. Some of what you have seen from your co-workers probably warms your heart and gives hope for your future work. However, not all of what you have seen felt right or reassuring. For example, perhaps you overheard some workers coolly discussing the murder of a child, which you recall from the morning paper. As the conversation went on, you realized that the child was a former client of the agency. Other than this conversation, it was business as usual, and few outward emotions were evident. If you did not know that such an outward appearance of insensitivity and callousness is one way workers cope with heinous acts of violence that are part of their workday, then you might have become very angry with the situation and very disillusioned about the people who work in the field.

It may be that you have seen some behavior for which there is no good explanation. Just as in any field, there are human service workers who are lazy, jaded, harsh, or unethical. Such circumstances can leave you feeling quite alone in your reactions. In addition, the support you need may not be forthcoming from the staff.

ISSUES WITH PEERS

Your peers are your classmates and other students who are interning at your site. These are the people who know best what you are experiencing. They know all about the feelings you share in class—and some of the ones you do not share! Peers can be an enormous source of support for you throughout your field experience. They can be there for you late into the night or early in the morning, across dormitories, and on telephones;

they can be there for you throughout the week when you need to talk with someone who really understands, and they are there for you in the seminar class when issues are raised and problems are addressed. On the other hand, they can also let you down by talking too much (or too little), offering unwanted advice, focusing only on their issues, or simply appearing indifferent.

One area in particular that can be very challenging is when you are working with your peers in the same agency. Same-site interns may be classmates on campus, in the same or different academic programs, or from different universities. Regardless, you may face some unique issues in these circumstances. The first is collegial rivalry, which can emanate either from the students themselves or from the staff. When this happens, students are compared and status is conferred, leaving one the loser and one the winner. A second issue is competition among the interns for attention, resources, support, or rewards. Competition in and of itself is not unhealthy or necessarily to be discouraged, as all can benefit from it. However, when competition becomes the focus of the energy in the field and each is watching the other's assignments, level of support, and staff involvement, difficulties can develop.

A third issue that creates challenges for same-site interns is differences in progress. There are lots of reasons why placements turn out as they do. However, when there are two interns at the same agency and they are having very different experiences, the peer relationship can be affected. No longer are the interns sharing similar experiences, and it is difficult to keep both involved in supporting each other. It seems that students whose internships are not going well benefit from a larger group of on-site interns. There will be less intense competitive pressure, and there is an opportunity for a broader base of support. Such differences in progress need to be addressed as potential issues by you, your supervisor, and your instructor.

The Organization and System

In Chapter 7, we discussed your placement as a system and a culture. By now, you have a better feel for that dimension of your work. You can sense it and you can see it. If not, you may want to review Chapter 7 and focus some of your energy on seeing these aspects of the organization. As with other aspects of your work, seeing these dimensions of organizational life unfold is often quite different from studying them or discussing them in abstract terms.

In particular, you should be getting familiar with the unwritten, informal rules and norms that are such a powerful part of life in any organization. What an agency values, for example, in its workers, its clients, and its work is evident in how supportive staff members are to each other, their commitments to the work, their attitudes about the clients, the work, the system, and their willingness to create a community. Learning to see this dimension of an organization can be fascinating. What you see, though, may not always reflect what you value or what the agency says it values, and you may find yourself in a value conflict with the agency and the system.

The paper versus people issue, for example, often becomes very real. There is rarely enough time to give clients all that they need and also attend to the myriad of forms, documentation, and procedures required by your agency and those agencies to which it is accountable. Sometimes something has to slide. "People before paper" may be what you value and it may be what the organization espouses, but you also see how much time is spent on paperwork, how important it seems that it be done just right, and how often an idea is put aside because of the paperwork it will create. Value conflicts such as this can be disheartening (Sparr, Gordon, Hickham, & Girard, 1988). What may surprise you even more is that in time, as you come to understand the pressures on the agency and the systems it interacts with, you might find yourself adapting to the very values that initially caused you concern.

Second, the agency and system have found ways to organize their resources to get the work done. As you look around your agency, you can see signs of the formal organizational structure, which is the way the agency works on paper. You see it in people's titles and in the ways people are divided to do the work. If you look even more closely, you will see signs of the informal organizational structure, or the structure of influence, which is what really happens in the course of the agency getting its work done (Stanton, 1981). Sometimes these informal networks function quite smoothly and support the overall goals and work of the agency. Other times, though, they seem to undermine those purposes. Although you have read about such discrepancies, actually seeing them operate and feeling their effects can leave you feeling confused at best and disoriented at worst. Yet this is how organizations tend to function for the most part. Ignoring this reality is sure to continually frustrate you in your work.

Third, you have a philosophy that is evolving through your internship. The agency has one, too, and here again it is the operational philosophy that counts, which you may now be starting to see. Many interns are impressed with their agency's operational philosophy. On the other hand, with the goals of efficiency and effectiveness tending to prevail in contemporary organizations and systems, there is also room for philosophical and political debate. You might find yourself frustrated by what you consider the coldness and slowness of the system, the reduction of people to inhuman status, and the realities of dealing with bureaucracies, as well as underfunded and understaffed community programs.

Self-Understanding

Finally, it is time to take a look at what you are discovering about yourself. You may have had some self-development goals in your learning contract. Even if you did not, the learning opportunities are there for you. Interns almost never fail to tell us that they learned a tremendous amount about themselves during their placements.

ASSESSING YOUR MEANING MAKING SYSTEM

One of the things we hope you are learning in the course of your placement is how to be reflective. Reflective workers in the helping professions think about what they do before they do it, while they are doing it, and after they have done it. Reflection, as you know, is an important part of the experiential learning cycle. However, Donald Schon (1983) maintains that reflective practitioners also reflect at a deeper level. Every time you make a decision, you take a variety of factors into consideration. Some of these factors arise only occasionally in your decisions, but others come up again and again. You may recognize these factors as part of your meaning making system as discussed in Chapter 2. Truly reflective practitioners take time to assess that system. They review their decisions, noting what factors went into them and what factors were ignored. You have made enough decisions in your placement that you can begin to form some ideas about your meaning making system. Be sure, though, to concentrate on what you have actually thought about and done, not what a book or professor advocates. In this case, the meaning making system you want to learn about is your own, not that of an author or instructor!

Parts of that system are the attitudes, values, behavior patterns, unresolved issues, psychosocial and cultural identity, and other aspects of yourself that you explored in Chapters 2 and 3. They are part of the filters through which you look at your life and make sense of your experiences. Often, these issues become much more alive as an internship progresses. Now is a good time to pause, perhaps review those chapters and your answers to the exercises, and think about what you have learned about yourself and your way of making meaning. Don't forget to include your learning style. Your earlier notions about how you learn may have been confirmed, contradicted, or some of both. If experiential learning is new for you, you may have discovered that your styles and preferences in this sort of learning situation are different from what you have experienced in the past.

BUMPING INTO YOUR OWN ISSUES

As in many other areas discussed in this chapter, many issues pertaining to yourself are not new. You wrote about them in Chapters 2 and 3 and may have thought a lot about them long before your internship. No one, however, can predict with complete accuracy how they are going to think, feel, and react until they get into a situation. And as with other issues, they often look different when they are actually touched and experienced than when they are anticipated or imagined. It is one thing, for example, to think about your values and what it may be like to have them tested or to be involved in a value conflict. If you have remained open to the human experience of the internship, however, you now know what it is like to have that happen.

Your experience at the internship can also teach you about aspects of yourself that you have missed up to now. You may not know you hold a value strongly, for example, until it is tested in some way. Or you may find yourself reacting strongly to a situation and not knowing why. After thinking about it and discussing it with your supervisor, you discover that the situation tapped an unresolved issue in your life.

Perhaps you have discovered or clarified some reaction patterns. You have probably found yourself in situations that are new and have learned something about how you respond. If you go back to Chapter 2, you may be able to state those patterns in the format encouraged there. If no patterns have emerged, try spending some time with your journal. Pick out several incidents or situations that stand out and try to analyze them using the three-column method. See if you can separate what happened from your thoughts in response and again from your feelings. Look for similarities across situations. If you see similarities in your feelings or thoughts, perhaps there are some similar conditions present in the events that happened.

FINDING MATCHES AND MISMATCHES

We have been encouraging you to take stock of your evolving relationships with your work, the people, and the system. However, we have not paid much attention to the other half of the relationship equation: you. If you have not done so, now is a good time to think a bit about what you have learned about yourself and how that learning helps make sense of the interpersonal experiences you are having. As you compare what you know about your clients, supervisor, co-workers, and other people involved with your placement with what you know about yourself, you can see where you are well matched to your internship and where you are not.

For example, one intern who worked at Planned Parenthood regularly fielded phone calls asking for information on abortion. Over time, she became impatient with the phone calls, the clients, and even the agency. She said that many of the questions she answered were "stupid" and wondered why the people were so poorly informed. She chafed at the bureaucratic procedures employed by the agency. Upon further exploration, it appeared that a values issue was at work. The intern was firmly opposed to abortion for herself and had been open about that belief with her instructor and the agency. However, she insisted that she believed with equal fervor that everyone should make their own choices and that she was not opposed to abortion in general. As the semester progressed, she grew more and more uncomfortable with the fact that abortions were performed at her agency; it seems that she had much stronger antiabortion beliefs than she thought. It took her some time to admit this change, even to herself, because she believed she should be tolerant. Once she did so, however, she was able to direct her anxiety away from clients and the agency and view it as a poor match, given what she had learned about herself. The agency arranged for her to work in their community education division, and the placement was ultimately successful.

Another intern was coleading a group with several residents at a residential facility. The other cofacilitator was also an intern, and they were regularly observed and supervised by an experienced staff member. The intern complained that his partner took up too much airtime, was constantly jumping in with opinions and suggestions, and didn't respect or want his contribution. The intern's partner did not see it that way, nor did the supervisor who came to observe. The intern was quite upset with both of them. However, as he looked at his journal and reflected on other situations in his life, he uncovered a dysfunctional pattern. When he was at all unsure of himself, especially in an academic situation, he had a very hard time speaking up, even though he often had valuable contributions to make. Time and time again, others would speak while he was still trying to muster the courage to do so. In this case, his partner was very verbal and thought quickly. Theirs was an unfortunate match. However, identifying it in this way helped them stop blaming each other.

These are just two examples. There are many other patterns and issues that might cause you some difficulty depending on the circumstances of your placement. Here are some more examples:

- You have dealt with the same issues as your clients, and you are not as "over it" as you thought.
- You have a hard time accepting criticism.
- You say "yes" when you want to say "no."
- You struggle to establish intimate relationships.
- You are extremely upset by confrontation.
- You have a need to smooth over conflict.

We encourage you to identify these and other issues that have arisen in your placement and to share them with you instructor, peers, and supervisor.

CONSIDERING YOUR LIFE CONTEXT

In Chapter 3, we encouraged you to think carefully about your life context, especially the other responsibilities you have besides the internship. Now, though, you have been actually doing the juggling act for a while. Furthermore, life may have thrown a few more balls in the air. You or someone close to you may have become ill, lost a job, been in an accident, and so on. How does it feel? Is your energy level still high or are you starting to bend under the strain?

We also asked you to assess your support system. Just as you cannot really know how emotionally demanding the internship will be, especially in combination with other aspects of your life, you cannot know how well your support system will function until it has been tested. On the other hand, the internship also offers some additional sources of mutual support. By working with your peers, you can broaden your perspectives and begin to develop your supervisory skills as you help them work on issues and they help you in return (Gordon & McBride, 1996). Many interns draw great professional and personal support from their supervisors, instructors, and co-workers.

Before reading the next section of this chapter, we encourage you to take half an hour or so and think about the first group of reflection questions (Part A) at the end of this chapter.

FACING REALITIES: THE DISILLUSIONMENT STAGE

As you worked through the last several sections of the chapter and the reflection questions, we hope you found many reasons to be happy. We also hope that you were able to think about logical next steps for yourself and some new or revised goals. However, chances are the exercises also highlighted some areas in which the internship may not be going the way you would like. Even if that did not happen, sooner or later you will face some problems at your internship. We are not talking about something minor that is quickly resolved. We are referring to situations that are more substantial, troubling, and that stay with you for a while. We cannot predict how many problems you will have, what they will be, or how you will react to them. Some interns experience only minor bumps in the road; others have a more profound period of upset and upheaval that causes them to question themselves and their placement. Whatever your experience is or will be, we hope this section helps you make sense of what is happening and guides you to approach it in the most growthful way possible.

Problems can happen anytime in an internship; perhaps you have already experienced some. If you are several weeks into the internship, it is a likely time for problems to emerge. For some of you, the initial glow of the internship has faded and some of the aspects that you do not like are becoming clearer. After you have settled in, people often expect more of you, and that can be hard. One common source of problems is the gap between anticipation and reality. There is almost always a difference between what you anticipated about the internship and what you actually experience in the field. Fortunately, if you acknowledged and dealt with your concerns in the anticipation stage, you will be less likely to have a wildly different reality from what you expected. However, there will be some discrepancies between your expectations—about clients, supervisors, co-workers, and yourself—and the realities of the placement, some of which may be troubling. It may be that the issues and personalities are not what you expected or that you are reacting differently than you thought you would. In addition, issues will arise in placement that you simply never considered or never knew existed because you had no way to anticipate them.

Problems can be small or large, but the first one you have usually has an added impact, precisely because it is the first one. You may not have expected to have problems, or at least not the ones you are now having. As we have pointed out many times, experiencing something is different than reading about or anticipating it. Once you have encountered—and resolved—the first problem or two, the experience will be different. Subsequent problems may be harder to resolve, but the shock of the first one or two will be behind you.

The sources of problems are as varied as interns and their placements. Some of the problems that our interns have encountered are listed below, written in the language of their student journals. Perhaps some of them look familiar to you. Perhaps you can add some of your own:

> I really can't accomplish much with these people [clients]—too much damage has been done and I can't perform miracles.
>
> I've never experienced what they have—they are writing me off. I can't work with them and they know it.
>
> A client lied to me. Or manipulated me. I thought they trusted me.
>
> A client had a relapse. Great. Now what am I supposed to do?
>
> There are some clients I just don't like. I can't find common ground with them.
>
> They just won't respond to me; not all of them but some of them.
>
> I have a good surface relationship but I can't move beyond that to really challenge them and explore some issues.
>
> The pace here is totally insane (or incredibly slow).
>
> My supervisor is too vague (or awfully blunt).
>
> My supervisor never seems to have time for me.
>
> My supervisor is always looking over my shoulder—she doesn't trust me.
>
> If I don't have any questions, my supervisor ends the supervision hour. That's not right. She's supposed to ask the questions.
>
> I don't like the way my supervisor treats the clients (or staff).
>
> The staff are cliquish—I don't belong.

They dump their busy work on me and send me on errands—now I'm a gofer!

They joke about clients behind their backs.

They are so cynical. BORING.

They are trying to get me involved in their problems. I'm just an intern here. I don't need to know what's going on. It's not my problem.

I'm just an intern. Why does everyone expect so much of me?

Just as there is a range of problems, there is a range of emotional reactions to them. Some interns seem to take the problems in stride; others are really thrown. As you read that sentence, try not to fall into the trap of saying to yourself, "Well, I am going to be the first type." You'll be very tolerant and supportive of those who do seem to be thrown, but you won't allow yourself that experience. Remember, allowing yourself to feel your problems, as well as to catalog and analyze them, is one way of remaining open to the human experience of the internship. There is no right way to react. Your reactions—and those of your peers—will depend on your emotional styles, the particular intrapersonal issues touched by a problem or situation, your willingness or unwillingness to be open with yourself and others about your feelings, and of course, the nature of the problems themselves.

In spite of our efforts to reassure them, interns often believe that if they talk about their difficulties or problems, or discuss their concerns or mistakes, then their grade in the field will reflect these shortcomings. It is true that the grade you earn in a field experience is influenced by your strengths as well as the competencies that need further development. However, your faculty instructor recognizes that all interns have shortcomings, as do practitioners and faculty. Your willingness to recognize and face the issues helps the faculty instructor gauge your growth in placement. In fact, not doing so may give the faculty reason to question how much you are gaining from your placement (Gordon & McBride, 1996).

Whatever Happened to My Internship?

For some interns, the problems they encounter are troubling aspects of an otherwise positive experience. Other interns reach a point where the whole emotional tone of the internship changes for them, and not for the better. Instead of enthusiasm and nervous excitement, they feel anger, resentment, confusion, frustration, and even panic. You may recall from Chapter 4 that this is called the Disillusionment stage.

If this description fits you, you are now coping with an enormous amount of the unexpected—not what you planned, but what you are living with during your internship. There is the unexpected end to the hopeful feelings of earlier days in placement and the onset of more negative feelings. There is the unexpected drop in your enthusiasm and the subsequent drop in your productivity. (Negative feelings do interfere with learning!) Concerns at this time center on many of the same areas as in the anticipation stage, except that there is a shift from the "what if?" concerns of anticipation to "what's wrong?" concerns: What's wrong with my internship? What's wrong with my clients? What's wrong with the organization? What's wrong with me?

Early on in your internship, we asked you to think about a metaphor that best described what your internship is like. Is that metaphor still working? How does your metaphor account for this period of diminished enthusiasm? We have often likened the internship experience to a roller coaster ride, with peaks and dips throughout, perhaps no different from the waves of highs and lows in your daily lives. However, this period is one of distinct character, a dip that leaves students, instructors, and site supervisors alike somewhat perplexed and overwhelmed at times.

The changes brought about by this shift in concerns are often subtle at first, but they persist and begin to become pervasive. You may be complaining to friends or family without even realizing it or hesitating before commenting on your internship to friends or co-workers. You may notice that you are having trouble getting to your placement on time, not looking forward to going, or muttering to yourself under your breath about the situation. Another unexpected consequence of this period is the tendency to direct your feelings at those around you, especially those who are connected to the internship experience—your site supervisor, clients, faculty instructor, co-workers, or even your peers. Sometimes the seminar class itself becomes the focus of your feelings.

The feelings can also be directed inward, and this can be a time when an intern's self-image takes a bit of a beating. For example, when clients do not improve as you expect, refuse to continue services, or react to you very negatively, it can call into question whether or not you were adequately prepared for the work, whether you are cut out to do the work, or whether you should be doing the work at this time in your education (Skovholt & Ronnestad, 1995). In addition, many students entering an internship are used to their friends telling them that they are very easy to talk with and helpful, and the students are often filled with a sincere desire to help. However, there is a difference between wanting to help and actually being helpful and between being an easy person for friends to talk with and being effective with challenging clients. When the inevitable difficulties and criticism come, these students may become resentful, perfunctory in their job performance, or critical of the supervisor (Blake & Peterman, 1985). Perhaps, too, you were not able to work collegially with co-workers or supervisors or within the agency guidelines; perhaps you found little compatibility of values or philosophy with the site.

Earlier in this chapter, we discussed the integration of theory and practice. Garvey and Vorsteg (1992) have suggested that this process can be quite difficult for some interns. It can be a time in placement when interns temporarily reject their previously held beliefs about the theories they work with and experience a crisis in confidence about the worthiness of the work they are doing. It is not unusual for these interns to become disoriented about their learning and appear to go into a funk, experiencing intense frustration, blaming their clients, and questioning their commitment to the work.

Generally, we have observed three potential categories for these feelings of disappointment and frustration. The first is *loss of focus* on the internship. Essentially, you are not doing what you went there to do because the focus of your placement has changed either intentionally or through neglect. Interns who lose the focus of their placements can feel it almost immediately; it is a feeling of being sidetracked or otherwise ignored or discounted. The second is a *loss of accomplishment*. You are not doing what you went there to do, not because the focus changed, but rather because you either are not able to demonstrate the skills or competencies needed to accomplish the internship or the design of your internship does not allow you to have the experience you wanted. You will probably feel frustrated and angry, feelings which you may direct at yourself or at the personnel or clients at your site. Finally, the third is a loss of meaning in the internship. Essentially, you are unhappy with what you are doing in placement. Either the work has no personal meaning for you, or the internship has not been designed to provide you with a meaningful field experience. Regardless, you may find yourself going through the motions of an internship, but without the spirit that comes from being personally invested in the learning and work.

There is no predictable time when these changes in feelings occur; for some interns, they never happen at all. Often, though, these changes are noticed during the phase of learning when the workload increases and there are greater demands for skills. However, the sense of disillusionment can occur earlier in placement, when anticipation concerns are highest, or later in placement, with concerns about competence in the work.

If you are in the middle of this sort of slump, or if you hit one later on, try to remember that it is perfectly normal; in our experience, more interns go through it than do not. Many authors feel that some level of dissonance is necessary for growth. Perry (1970) argues that shifts in ways of thinking about the world are caused by conflict and dissonance. Similarly, Langer (1969) identifies conflict as playing an important role in progress and development and contends that for change to happen, you need to feel that something is wrong. He calls this a state of disequilibrium and says that it is necessary to raise the energy and emotions you need to make changes and put your internship back on track. Furthermore, many authors and researchers note that a period of disillusionment is a normal part of any internship (Garvey & Vorsteg, 1992; Lamb et al., 1982; Suelzle & Borzak, 1981; Sweitzer & King, 1994).

SUMMARY

This time in your internship is full of opportunities. In the early part of this chapter, we discussed the opportunity you have to keep pushing yourself and to keep your internship fresh and challenging. Problems, on the other hand, may seem more like a curse than an opportunity. We certainly feel that way sometimes. However, you will handle your problems better and learn more from them if you take a more constructive view.

Whether you are experiencing isolated problems or a more pervasive disappointment in your internship, facing and resolving the issues confronting you are very important. You might be able to ignore the problems, but chances are that will make things worse. Making sense of the feelings and concerns of this stage in an internship is critical to growing through it. And growing through it is critical to a successful internship. Growth lies in the constructive activity of putting things back in order; resolving the conflict is a prerequisite for progress to the next stage of development and for normal and healthy adjustment to occur. Remember, the new state of equilibrium (feeling that all is well) does not cause progress to occur; rather disequilibrium is the source of growth and development (Langer, 1969).

An interesting and useful perspective on problems and their resolution is offered by Watzlawick, Weaklund, and Fisch (1974). The terms *problem* and *difficulty* are often used interchangeably, and indeed we have done so in this book, but Watzlawick et al. draw a distinction between them. Difficulties, they say, are normal, if unpleasant, events that either have a fairly simple solution or no solution at all. However, in our attempts to solve these difficulties, we often make them worse or bring on additional troubles. Then say Watzlawick et al., we have a problem. Many of the issues you face as an intern are normal and common. Knowing that may make it a little easier for you, but probably not much. However, if you ignore or misdirect your energies in trying to address difficulties, they can snowball into full-blown problems.

You may recall that in Chapter 4 we described the Disillusionment stage as a crisis of growth and a time of risk and opportunity. If you do not try to face and cope with the problems that arise, you can end up feeling stuck. Those feelings of resignation and resentment will not go away; instead you will suffer bravely (or not so bravely) through them, just as you would a course that turned out to be a disappointment. On the other hand you have an opportunity to learn more about your work and yourself. More important, you have an opportunity to feel empowered. If you can get through this time, you will feel a new kind of confidence. This is not the false bravado or naiveté that tells you there won't be any more problems, but the confidence that you can resolve them and learn about yourself and others in the process.

Some interns at this point in the placement begin to question their choice of career and their suitability for it. That can be quite unnerving, especially so close to the end of your professional preparation! Questioning your choice of career may make sense, though, and not only because affirmation of the choice is very important. It also makes sense that the choice can withstand the scrutiny of reflection at any time, especially at the beginning of your career, just before you commit yourself to your first position. You may in fact conclude that the specific career you had your eye on is not for you after all; it does happen. Finding that out now, and not in the middle of your first job, is a blessing for all concerned. Furthermore, if you have come this far, you have certainly gained knowledge and skills that are going to be useful to you whatever you do. However, this is no small problem you are facing, and it will need to be dealt with, too.

The first step in moving through the challenges that face you is taking a clear inventory of them. Earlier, we asked you to consider how you were progressing in several areas. Perhaps some issues surfaced then. Perhaps others have crystallized as you read this section. The second step, of course, is developing a plan to address these issues, and that is the subject of the next chapter.

For Further Reflection

PART A: GROWTH AND PROGRESS

1. Reread your answers to the Anxieties and Anticipations exercise at the end of Chapter 5. Have your concerns changed over these first weeks? If so, how?
2. Go back and reread your daily journal entries. What would you say are the major things you have learned so far about your clients, about working with them, the agency, and about yourself?

3. When it is time to go to your internship, how do you usually feel? Be honest. Has that changed over the course of the semester? If so, what do you make of that change?
4. Take a look at the goals in your learning contract. List them in your journal and indicate whether each has been met, not met, or partially met. In the latter cases, are you disappointed? How might you move beyond these goals to the next logical challenge?
5. What things have happened that you didn't expect? Include positive and negative things and think about the three areas (knowledge, skills, and self) as well as your responsibilities, the client's, and your supervisor's.
6. In general, would you say the tasks you have been given so far have overchallenged you, underchallenged you, or been about right? Explain your answer.
7. Are there clients you are beginning to feel close to? Are there some you have a hard time connecting with? What are your ideas about why?
8. What specific issues do you see for specific clients that go beyond the basic establishment of a relationship with you?
9. What sorts of challenges have clients presented you with? Have they tested your boundaries? In what way? What other boundary-related issues have surfaced?
10. What have you learned about your supervisor's style and how well it is or is not matched to your needs?
11. Have you and your supervisor disagreed yet? If so, how did that happen? How did it make you feel? How was it resolved?
12. Have supervisors or others given you verbal feedback on your performance? How did you feel about what they said?
13. How accepted do you feel by your co-workers?
14. In what ways have your co-workers met and not met your expectations of them?
15. Do you have any concerns about your relationships with your peers—other interns at your site and/or students in your seminar class?
16. How has the agency itself compared to your expectations? Does the pace approximate what you thought it would be? What are the unwritten rules, or values, of the agency? How do you feel about them?
17. Look back at the questions you answered at the end of Chapter 2. Are any of these, or any other, intrapersonal issues coming up at the internship? How are they affecting you?
18. How are you doing with managing your other responsibilities? What are the areas of stress or friction?
19. Now that you have been at your placement for a while, does the support system you described in an earlier entry seem adequate? If not, in what categories is it failing you?

PART B: FACING REALITIES

1. In light of the work you have just done, are there some new goals you want to set for yourself? Be specific.

2. What problems have you become aware of at your internship? Make a list of them and try to put them in order of importance.

For Further Exploration

Albert, G. (Ed.). (1994). *Service learning reader: Reflection and perspective on service.* Raleigh, NC: National Society for Experiential Education.

This outstanding collection of readings was compiled by the staff of the Center for Service Learning at the University of Vermont and is used as a textbook in their field study program. The articles are inspiring and informative, and they offer varied perspectives on service to others and on self-understanding.

Benner, P. (1984). *From novice to expert.* Menlo Park, CA: Addison-Wesley.

This seminal book in clinical practice in nursing looks at five levels of competencies and the demonstration of excellence in actual practice. It is particularly valuable because of the dialogue with nurses on which it is based and its applicability to human service work.

Borzak, L. (Ed.). (1981). *Social work field instruction: The undergraduate experience.* Beverly Hills, CA: Sage Publications.

This text is a compilation of articles that includes students' perceptions of the field experience. The articles are enlightening and offer students an understanding of how other students experience their fieldwork.

Dreyfus, S. E., & Dreyfus, H. L. (1980). A *five-stage model of the mental activities involved in directed skill acquisition* (Unpublished Report F49620-79-C-0063). Air Force Office of Scientific Research (AFSC), University of California, Berkeley.

This brief, yet informative, paper written by a mathematician and systems analyst and his brother, a philosopher, describes their model of skills acquisition, which is based on a study of chess players and airline pilots.

Garvin, D. S. (1991). Barriers and gateways to learning. In C. R. Christenson, D. A. Garvin, & A. Sweet (Eds.), *Education for judgment* (pp. 3–13). Boston: Harvard Business School Press.

The power of dialogue is discussed as part of the process of learning, self-discovery, and self-management.

Ronnestadt, M. H., & Skovholt, T. M. (1991). A model of professional development and stagnation of therapists and counselors. *Journal of the Norwegian Psychological Association*, 28, 555–567.

This team of writer and therapist developed a model explaining periods of growth and nongrowth in the counselor's professional development.

Rosenthal, R., & Jackson, L. (1968). *Pygmalion in the classroom.* New York: Holt, Rinehart & Winston.

A classic text, illustrating the power of expectations on performance.

References

Benner, P. (1984). *From novice to expert.* Menlo Park, CA: Addison-Wesley.

Blake, B., & Peterman, P. J. (1985). *Social work field instruction: The undergraduate experience.* New York: University Press of America.

Brill, N. (1998). *Working with people: The helping process.* (5th ed.). New York: Longman.

Burnham, C. (1981). Being there: A student perspective on field study. In L. Borzak (Ed.), *Field study: A source book for experiential learning* (pp. 65–72). Beverly Hills, CA: Sage Publications.

Deutsch, C. J. (1984). Self-reported sources of stress among psychotherapists. *Professional Psychology: Research and Practice, 15*(6), 833–845.

Dreyfus, S. E., & Dreyfus, H. L. (1980). *A five-stage model of the mental activities involved in directed skill acquisition* (Unpublished Report F49620-79-C-0063). Air Force Office of Scientific Research (AFSC), University of California, Berkeley.

Faber, B. A. (1983). Psychotherapists' perceptions of stressful patient behavior. *Professional Psychology: Research and Practice, 14*(5), 697–705.

Garvey, D. C., & Vorsteg, A.C. (1992). From theory to practice for college student interns: A stage theory approach. *The Journal of Experiential Education, 15*(2), 40–43.

Georges, R. A., & Jones, M. O. (1980). *People studying people.* Berkeley: University of California Press.

Gordon, G. R., & McBride, R. B. (1996). *Criminal justice internships: Theory into practice.* Cincinnati, OH: Anderson Publishing.

Kegan, R. (1982). *The evolving self: Problem and process in human development.* Cambridge, MA: Harvard University Press.

King, M. A., & Uzan, S. L. (1990). *The field experience: An integrative model.* Workshop presented at the annual conference of the National Organization for Human Service Education, Boston.

Lamb, D. H., Baker, J. M., Jennings, M. L., & Yarris, E. (1982). Passages of an internship in professional psychology. *Professional Psychology, 13*(5), 661–669.

Langer, J. (1969). Disequilibrium as a source of development. In P. Mussen, J. Langer, & M. Covington (Eds.), *Trends and issues in developmental psychology* (pp. 23–37). New York: Holt, Rinehart & Winston.

Perry, W. G. (1970). *Forms of intellectual and ethical development.* New York: Holt, Rinehart & Winston.

Royse, D., Dhooper, S. S., & Rompf, E. L. (1996). *Field instruction : A guide for social work students* (2nd ed.). New York: Longman.

Schon, D. A. (1983). *The reflective practitioner: How professionals think in action.* New York: Basic Books.

Skovholt, T. M., & Ronnestad, M. H. (1995). *The evolving professional self: Stages and themes in therapist and counselor development.* New York: Wiley.

Sparr, L. F., Gordon, G. H., Hickham, D. H., & Girard, D. E. (1988). The doctor-patient relationship during medical internship: The evolution of dissatisfaction. *Social Science Medicine, 26*(11), 1095–1101.

Stanton, T. K. (1981). Discovering the ecology of human organizations. In L. Borzak (Ed.), *Field study: A source book for experiential learning* (pp. 208–225). Beverly Hills, CA: Sage Publications.

Suezle, M., & Borzak, L. (1981). Stages of fieldwork. In L. Borzak (Ed.), *Field study: A source book for experiential learning* (pp. 136–150). Beverly Hills, CA: Sage Publications.

Sweitzer, H. F., & King, M. A. (1994). Stages of an internship: An organizing framework. *Human Service Education, 14*(1), 25–38.

Wagner, J. (1981). Field study as a state of mind. In L. Borzak (Ed.), *Field study: A source book for experiential learning* (pp. 18–49). Beverly Hills, CA: Sage Publications.

Watzlawick, P., Weaklund, J. H., & Fisch, F. (1974). *Change: Principles of problem formation and problem resolution.* New York: Norton.

Wilson, S. J. (1981). *Field instruction: Techniques for supervisors.* New York: Free Press.

Yuen, H. K. (1990). Fieldwork students under stress. *American Journal of Occupational Therapy, 44*(1), 80–81.

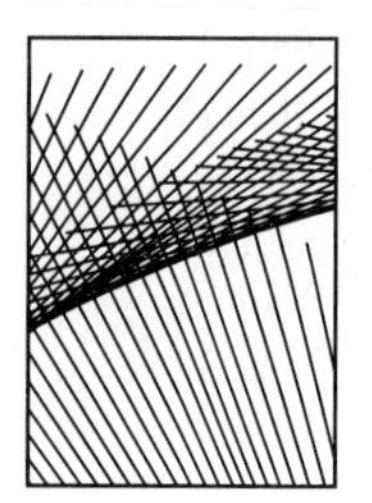

CHAPTER 9

Breaking Through Barriers: The Confrontation Stage

If students don't have a chance to address their fears by having the chance to prove themselves, then at graduation they will walk away with the same fears instead of having overcome them.
STUDENT JOURNAL ENTRY

In the middle of difficulty lies opportunity.
ALBERT EINSTEIN

After working your way through the exercises in the last two chapters, you have a clearer and more concise understanding of the issues, both large and small, that you are facing in your internship. If you think about it, though, you will realize that you have already managed to prevent a number of difficulties from turning into problems. We have been encouraging you to focus on those that are still unresolved, but don't lose sight of your previous victories—however small they may seem to you now—as you prepare to confront those that await you. This is the confrontation stage, an opportunity for you to resolve some troubling issues, to gain confidence in yourself, and to continue to grow.

THE TASKS YOU FACE

The confrontation stage can be the most challenging time in your placement. It demands that you take charge of your internship in ways that you may not yet have done. The process of resolving the issues that continue to challenge you tends to take center stage. As you struggle to achieve a sense of independence, confidence, and effectiveness in your work (Lacoursiere, 1980), you are faced with the reality that such qualities are born not just of your skills and accomplishments but of your ability to overcome obstacles. Breaking through the barriers, for now, is as important as the work in your internship.

We are going to offer you a model for approaching the problems you choose to take on. It will combine easily with any other models and approaches you may know. You may, after reading it, find that there is one you like better; that is fine with us. Regardless of the one you choose, though, a model itself is not going to be enough. Change can be very difficult. Perhaps you have had this experience either in or outside of your internship. You identify a problem and think you know clearly what is involved. You really want to do something about it. But every time you try to think clearly, you become confused, lethargic, and drained until you finally start to think about something else just to get away from those feelings. Or you may find that you begin making a little progress, only to drown in the frustration and anger that you feel about the problem and your perceived inability to do something about it. This sort of paralysis is extremely frustrating.

You may feel like you don't know what to do, but knowing is not the problem. And that is the limitation of even the best models. They are largely cognitive; they involve your thoughts more than your feelings. We have found that no model is going to work without three important ingredients—belief, will, and effort—which, interestingly, come not from your thoughts, but rather from your heart. The affective domain must be an equal partner with the cognitive if you are to be successful in confronting the problems you have identified.

The Power of Belief

If you do not believe that a situation can change, most likely it will not. Let's rephrase that. If you wait for a situation to change on its own, it might. However, the point of this chapter is to help you be an active agent of change. If you do not believe that you really can change a situation, then you probably cannot. Your perceived ineffectiveness will became a self-fulfilling prophecy. Gerald and Marianne Corey have written a marvelous book on self-change and effectiveness entitled *I Never Knew I Had a Choice* (Corey & Corey, 1997). Perhaps *I Never Believed I Had a Choice* more accurately reflects the reader's experience. You need to believe that things can be different and that you can create the change that needs to take place.

Of course, there is a somewhat vicious circle here. You can work on feeling confident, and that will help you succeed, or you can work on solving the problem, and that will help you feel more confident. For some people in some situations, the belief is the foundation for the action. For other people in other situations, the converse is true. You know best what works for you. However, if you begin to try to effect a change and feel your confidence falter, you may want to step back and see where your lack of confidence is coming from (a review of Chapter 2 may help you here).

The Power of Will

Will is the determination to make change occur. The frustration and anger you feel are normal parts of the emotional experience of trying to change something. You need to allow yourself to feel those feelings and not run from them or hide them. As with most feelings, that is how you begin to work through them and allow determination to come to the fore.

Another barrier to this determination is the resentment some interns feel about having to work on the change. You may feel that this problem is not of your making, and you may be right. So if you didn't "do it," then why should it be you who has to fix it? Accepting responsibility can feel like you are accepting blame for the situation. But you are not accepting blame; you are empowering yourself. As long as you wait for others to act different, you are not in charge. Change may come, and you may enjoy it, but what will you do the next time? Wait and hope that you luck out again? We are trying to empower you, to help you feel able to take charge of your life and your situation when you want to.

The Power of Effort

Finally, you have to do something. Effort refers to what you actually do. It also refers to persistence and perseverance. Some of your difficulties will be relatively easy to analyze and resolve. Others take more time.

Belief, will, and effort are, of course, interrelated. You can probably see elements of all of them in the description of each one. And some of you may be feeling impatient. "Enough," you may be thinking. "I am ready. I am willing. Let's go!"

It will not take long to read the model that follows, but it takes focus and practice to use it well. We hope you are willing to make the effort and that you find the effort worthwhile. But if you are struggling with it and you are sure you understand it, you may want to come back to this section and consider these issues again.

A METAMODEL FOR CHANGE: EIGHT STEPS TO BREAKING THROUGH BARRIERS[1]

The literature is replete with techniques for solving various kinds of problems and with generic problem-solving models (see, e.g., McClam & Woodside, 1994). Most models include steps that answer the following basic questions: (a) What is the problem? (b) What do I know about the problem so far? (c) What is my goal? (d) What are my alternatives for reaching that goal? (e) What is my plan of action? This generic framework is hardly a new idea; in fact, it has its roots in the ancient teachings of Buddha, who wrote about the Noble Truths, including suffering, the origins of suffering, and the path leading to cessation of suffering (Watzlawick, Weaklund, & Fisch, 1974). Although our model is hardly as profound as its prototype, it does include these basic steps, with a couple of twists to traditional problem solving.

First, the model is designed to slow you down a bit. Yes, you read that correctly. If you have a problem that is really troubling, you are probably focused on what to do about it, what action to take. Anything, you figure, has got to be better than this. Or you

---

[1]This model is derived from a model for interdisciplinary integration and problem solving (see Sweitzer, 1989) that in turn has its roots in the Behavior-Person-Environment (BPE) principle advanced by Kurt Lewin (1954).

choose a strategy because it worked in another situation or because a friend used it successfully. Thinking intelligently about a problem involves some careful analysis, some goal setting, and finally, deciding on and implementing a plan of action.

This model is also designed to broaden the way you think about your problem. We want you to move beyond the initial stage, where you hold one individual responsible, and look at the perspectives and contributions of many other people. We also want you to consider the wider context in which the problem is occurring, including the systems that are involved, such as work groups, the office as a whole, and other agencies with whom you work. This model asks you to think and feel your way through a process, not simply to plug in answers to questions and in turn get a score that instructs you what to do next. It also involves a process that you may want to engage in with others. For example, you may want to work with other interns at your site, or you may want to discuss your issue(s) in seminar class using the model for change.

We suggest that you first read through the model in its entirety and follow the example we have included. Next, return to the first step and, using an issue that is problematic to your internship, work your way through all eight steps. To help you through the steps, we have provided guiding questions and identified potential pitfalls. After you complete the tasks of each step, you will be asked to write a brief paragraph responding to the guiding questions.

Eight Steps to Creating Change

The mere formulation of a problem is often more important than the solution.
ALBERT EINSTEIN

STEP 1: SAY IT OUT LOUD

Quick—in one sentence—what is the problem? Don't think too much; in the next step, we will let you name the problem more carefully. Just say it and then write down what you said. You can learn something important about how you are thinking from the way you first state the problem. Consider an intern who is struggling with a client. Here are some ways the intern might initially describe it:

- I can't get anywhere with her.
- She is not interested in changing.
- This client needs a different program.
- No one told me how to handle this.
- Their approach isn't working with this client.

Do you see how each of these statements locates the problem in a different place? Where did you initially locate your problem? Where are some other places it might be located?

STEP 2: NAME THE PROBLEM

Now it is time to let go of the feelings and the impulses—the name calling and blaming—and think clearly about the behaviors that are causing problems for you. Don't think about the people; think about the specific behaviors. Behaviors are actions that anyone can see if they are able to observe. You also need to think about the thoughts and feelings you are having as a result of these behaviors.

Here is one way to do that using a process that will be somewhat familiar to you if you have been using the "three-column" method in your journal entries. Take a piece of paper and divide it into three columns. In one, list the behaviors. In the next column, list your thoughts that result from those behaviors. Finally, list the feelings that result from each behavior. Now, look at what you have written and try to state the problem in one sentence. It can be a long sentence, but try to use just one. Think about when and in response to what this problem usually occurs and see if you can express that in a sentence. If it is not a recurring problem, this sentence will not be important.

In our experience, though, many of the problems that interns choose to confront in this stage are issues that have come up again and again, although they may appear different because the specific problems they create can vary. For example, an intern who has issues with acceptance may choose to eat lunch alone and thus not be included in the energy and discussion that go on in agencies where working lunches are part of the normal routine. Problems can develop if the staff interprets the intern's choice to eat alone as an indicator of a lack of investment or interest in the issues discussed at lunchtime. In the next sentence, see if you can describe exactly how this problem affects you. In what specific way does it create unpleasantness? Does it embarrass you? Does it make you angry? Does it hamper your ability to work?

Once you have worked on this group of sentences, you will have a concise, clear statement of the problem. These sentences comprise a paragraph. Add it to your worksheet. Here is an example:

> When she runs away from the program, it angers me because I think about all the work I have done with her, and she just blows it every time. It's just like when I used to work with kids on the streets. They would do the same thing, not listen to me. It makes me feel helpless, sort of powerless, and useless in my work.

STEP 3: EXPAND YOUR THINKING

This step is not unlike the reflective observation step of Kolb's learning cycle (Kolb, 1984). You are asked to think about the problem by examining its components from different perspectives and along different dimensions. Although taking perspective and examining it are insightful experiences, they can also be frustrating for reasons we have already discussed. But someone has to do the investigative legwork, and in this case, the only one who can is you. Begin with some scrap paper. As you work your way through the following four guiding questions, try to keep in mind the role of politics and other subtle dynamics. Be patient and persistent with yourself; some of these steps may not come easily to you, but they all can be mastered with a little perseverance.

Who are the players involved in this issue? Identify as many people as possible who are related to the problem. Including them does not mean that they are causing or contributing to the problem, only that they are part of the scene. Don't forget to include yourself.

How do each of them see the issue? Work your way through each person on your list and try to put yourself in that person's shoes. Try to see the situation from their points of view by imagining what it is like to have their positions on the staff, to have been there as long as they have, to work with an intern, or to have you in their work life. Some of our students have found it helpful to imagine all these individuals forming a circle, which in turn forms the hub of a wheel. Each person on the wheel sees the issue differently depending on where they are located. Their perspective, in turn, influences their stand on the issue at hand. Try to imagine how each of these individuals views and feels about the problem. From your perspective, how might each of them be contributing to the problem? How about from their perspectives?

What are the major systems involved in the situation? Each of these individuals is part of and influenced by a number of systems. (You may want to review Chapter 7 if the notion of systems is not yet clear to you.) Your first task here is to make a list of all the systems that are somehow connected to the problem. You may find it easier to approach this task by brainstorming. List all the possible systems you can think of. Begin with those directly connected to the situation and move to those with indirect connections. Or you might find it easier to be more organized and work your way from the inside out or the outside in. The inside group of systems includes the staff (and any subgroups of the staff you can identify), your work group, the clients, and anyone else who connects directly with the agency. Now think about the agencies that work with yours, either collaboratively, competitively, or adversarily. Next identify the sources of funding and other resources. Don't forget systems on campus, including your professors, the administration, and the seminar class. There may be more; if so, keep going! Now go back to your list and eliminate any systems that, on further consideration, seem to have very little to do with the problem.

Now think about how each system—as opposed to the individuals in it—contributes to the problem. To do this, you need to focus on the rules and roles that drive individual behavior. Again, you may want to review this section of Chapter 7.

What about you? Yes, you! In all likelihood, you are contributing to the problem in some way. Once again, we urge you to return to an earlier part of the book, this time to Chapters 2 and 3 and the exercises you did. Are some of your personal issues and values being touched by this situation? Your initial statement of the problem (Step 1) may be a clue here. Does your response to this situation remind you of any other situation or group of situations? Is there a pattern or a set of irrational beliefs at work?

Now go back over all the ideas and insights you have had during this portion of the problem analysis. Write a paragraph discussing them. Notice how you have begun to create a new picture of the problem that is more complex and takes into consideration multiple perspectives on the issue.

STEP 4: CONSIDER THE CAUSES

Now go back over your notes and summary from the last step and try to come up with a list of the major causes of the problem. Notice we did not say the major cause; there is always more than one. Be sure to consider all the individuals and systems you examined before concluding your list.

STEP 5: FOCUS YOUR ATTENTION

You have identified a number of dimensions of the problem. You will need to make changes and alter the dynamics discussed in the cause statement, but where do you begin? First, try dividing a piece of paper in half and labeling the columns "what I can change" and "what I cannot change." We are asking you to think about the issues over which you do and do not have influence. Let's look at these lists more carefully, this time in terms of what you can and cannot do about the issues. When you do have control over the issue, the responsibility is to do something about it; that is, do what you can do about it. When you have no control over an issue, your responsibility is to accept it; that is, resign yourself to accepting this as something that you must live with.

The possible pitfall in this commonly used exercise is the tendency to confuse what you can do something about with what you cannot do anything about. As you look over the lists, check for those issues that you continually try to do something about but actually have no control over. Being caught in this danger zone is like spinning the wheels of your car on ice: You are going nowhere fast and using up a lot of energy trying to get there! There are also those issues that you have accepted but can actually have some control over and do something about.

It is time to go back and add a new paragraph to the statement you are developing. This paragraph needs to answer the questions: What is blocking me from resolving this problem? What is it that I need to change?

STEP 6: DETERMINE YOUR GOALS

For each cause of the problem that is on the "changeable" side of your list, you now need to develop goals for change. State as specifically and behaviorally as you can what you want to be different. There is no room for fuzzy thinking here. Just wanting the issue to go away may be your most salient emotion, but now is the time for clear and concrete statements. Be sure that you have covered all the individuals and systems that you believe have a substantial role in causing the problem.

One possible pitfall here is to confuse a goal with an action step. For example, suppose that one major cause you identified is that your duties are radically different from what you expected and were told. Changing placements is not a goal; it is an action step. The goal would be to bring your daily actions more in line with your stated learning goals. One reason this is important is that there are usually multiple ways of achieving any goal. If one doesn't work, take some time to be disappointed, but then think of another. In the example just given, if you made switching placements your action step and then were told you could not make the change, you might feel the problem is now unsolvable. If you keep your real goal in mind, though, you are more likely to come up with another approach.

STEP 7: IDENTIFY THE STRATEGIES

Your task here is to develop a list of possible concrete actions you could take to meet each of the goals you identified previously. In developing the action steps for each goal, be sure the interventions reflect your knowledge about the nature, cause, and context of the problem. The principle of effectiveness is operating here; that is, each action must be effective for each goal. In other words, the action steps you choose must allow the goal to be realized. The picture you have created of the problem in this step is one that is action packed and future focused. When you write your paragraph about this step, let yourself be guided by the question: What must I do to make my goals happen?

You may remember that we called this model a metamodel because it allows you to incorporate many other approaches and strategies. You have probably studied many approaches to resolving interpersonal problems and conflicts, making constructive changes in yourself, and making change at the system or community level. But if you have not or if you want to do some more exploration before deciding on a strategy, we have included some resources at the end of the chapter.

STEP 8: CREATE THE CHANGE

This is the step you have been waiting for! Ever since the issue became a barrier to your internship, you have focused on this step. It is the time when change happens. If you've done your work with determination and effort, with genuineness and a willingness to be reflective and introspective in your thinking, and if you asked yourself the tough questions that needed to be asked, then you are ready to move on. It is time to do the achievable and implement your plan with commitment and perseverance. Yes, you've given it your best, and your best is good enough! The potential pitfall of this step is to fail to do what you can do about the issue, resigning yourself to the status quo when you have the capacity to make change occur.

The guiding question for this step is: How will I know when things are different? Let this question guide you as you implement your plan and move through the barrier that has prevented you from moving ahead in your internship. Let it also guide you as you write the last paragraph of your problem statement. Congratulations. You did it!

Not Being Taken Seriously: The Case of the Rebuffed Intern

Some of you may be comfortable with this model and can work with it right away. But for others, it may be too abstract. So we are going to work through an example for you. We will not detail all aspects of the steps, but we will show you how the eight steps work in this particular case.

Leslie is an administrative intern working at a group home that is undergoing a major licensing review. The site supervisor is an administrator. One of the tasks of Leslie's internship is to work with the supervisor in preparation for the review, a task which seemed particularly exciting. However, several weeks into the internship, Leslie was very unhappy and ready to switch internships. Here is the problem as this intern worked through it, written in the style of a series of journal entries.

STEP 1: SAY IT OUT LOUD

Once able to talk about the problem, Leslie said, "They don't take me seriously!" Just saying the words brought to the surface feelings that had been bubbling below for quite a while.

STEP 2: NAME THE PROBLEM

I was told I would have a major role in preparing for the visit. My supervisor said he would help me understand the regulations, let me do some of the research and even some of the writing, and meet with me regularly to discuss my work. Really, none of this has happened. There is always an excuse and a promise to make it up to me later. I am bored because I don't have anything to do that is worthwhile and makes me think; and I am frustrated because there is important work going on around me and I am not included in it. I'm beginning to think they don't trust me. I spend a lot of time trying to think of what I could have done wrong. I just don't know.

STEP 3: EXPAND YOUR THINKING

My supervisor I guess he is under tremendous stress about this on-site review. I suppose he feels responsible for how things turn out for the agency. He is pressed for time. Bet he feels it may be faster to do it himself. I bet he is afraid of his boss. Interesting.

My co-workers For whatever reason, they don't talk with me about the visit. I see that they are very busy. I know that they had bad experiences with the last intern. But they don't know my skills. Guess they haven't had a chance to see them.

The entire staff I know that there have been threatened staff cuts. I think somehow that the possibility of cuts creates a "rule" that everyone must make themselves seem as vital and indispensable as possible. That's all that I see—everyone acting important. Must have something to do with the possibility of cuts.

The facility I know that this review visit is very important to the facility's future. Exactly how, I am not yet sure. I've heard the staff talking about the possibility of declining referrals if there is a problem with the visit. That means their jobs, I guess.

My issues I know that I am very sensitive to being put off. I really have problems when that happens. I know that it is my issue because other students in seminar class have had similar situations, but they did not make issues out of them. I remember hearing them talking about their situations. I cringed, but they did not. Another thing I realize is that it is hard for me to be persistent. I actually wonder if I can do this, and I'm terrified I might make a mistake. Just the idea of having to continue to deal with this scares me. You know, I do fine when I have lots of support and attention, but I wilt quickly when I feel the least bit slighted. It really doesn't take much.

STEP 4: CONSIDER THE CAUSES

1. I don't have enough faith in myself—I just don't believe in myself at times.
2. I give up too easily!
3. My supervisor is not living up to our agreement.
4. Staff members don't understand my role or what I am able to do.

STEP 5: FOCUS YOUR ATTENTION

I can't change the pressure that the site visit is causing on my supervisor and on the rest of the staff. I cannot change that this is my internship site (even though I want to at times, but I guess that the field coordinator on campus and my campus supervisor would say, "No—you have got to work this out!"). However, I can change all the causes listed in Step 4.

STEP 6: DETERMINE YOUR GOALS

I can think of three primary things that have to happen in order to make this internship work for me. I must:

1. *Increase my assertiveness and self-confidence.* If I do this, I will feel good about myself even if I am rebuffed by others, especially those whose opinions matter a lot to me. After all, not everyone cringes when they feel rebuffed.
2. *Get my supervisor to honor our agreement.* It cannot be that difficult. After all, it is important to him to do a good job, and he does take pride in having an intern. If I can get him to realize that an agreement is an agreement, and that both I and my faculty supervisor believe that I can do the work, things will change for the better. Then I will have something worthwhile to do.
3. *Make sure the staff knows why I am here and what I can do.* I have got to become more visible and better known in the office. Once the staff starts to get to know me, they will want to know why I am in their work space. If I can do that, I will have succeeded in getting them to know me and my work.

STEP 7: IDENTIFY THE STRATEGIES

1. I can learn a lot more about myself if I:
   - use the trumpet model to work on being able to persist.
   - use Ellis's ABC model to understand why I wilt so quickly.
2. I can get my supervisor to honor our agreement and the contract if I:
   - request a special meeting.
   - write my concerns down ahead of time.
   - get support from my faculty supervisor.
3. I can make sure the staff gets to know why I am here and what I can do if I:
   - ask my supervisor to invite me to a meeting and introduce me.
   - volunteer to help the staff with some of their regular duties.

STEP 8: CREATE THE CHANGE

Believing I can bring about change, I must persevere and continue until I see the changes that suggest:

1. I am feeling more aware of my role in bringing about change and more confident about my work and the future of my internship.
2. My supervisor is more aware of how the internship is affecting me and is willing to reconsider the learning contract we created in order to better meet my needs.
3. Staff members are more invested in their relationships with me and are actively reaching out to involve me more in their work.

I guess I am surprised at what I can do to make my internship different. It really is not all that difficult. Makes me wonder why I have been stuck for so long.

SUMMARY

This chapter has tried to get you ready to confront the issues you have identified and given you some tools to do that. However, sometimes it is also helpful to pay attention to how others approach and resolve problems. Look around you. You probably won't have to look far until you find someone—a friend or a peer—who thinks very differently than you about problems and their solutions. What are those differences? How might these people have dealt with the issue(s) you are facing? Can you learn anything from their approach?

For Further Exploration

ON PROBLEM SOLVING

McClam, T., & Woodside, M. (1994). *Problem solving in the helping professions.* Pacific Grove, CA: Brooks/Cole.

Clearly written and concise with a problem-solving model and lots of cases to study.

ON INTERPERSONAL EFFECTIVENESS

Corey, G., & Corey, M. (1997). *I never knew I had a choice* (6th ed.). Pacific Grove, CA: Brooks/Cole.

Excellent book with a good chapter on relationships.

Johnson, D. W. (1990). *Reaching out: Interpersonal effectiveness and self actualization* (4th ed.). Upper Saddle River, NJ: Prentice Hall.

Clear and incisive with lots of examples and exercises—a classic.

ON MAKING INTRAPERSONAL CHANGE

Brill, N. (1998). *Working with people: The helping process* (5th ed.). New York: Longman.

Self-understanding and self-modification are discussed throughout this book.

Ellis, A., & Harper, R. A. (1975). *A new guide to rational living.* Upper Saddle River, NJ: Prentice Hall.

Ellis's ABC model is explained with helpful self-help guides to work through your own situations.

Sweitzer, H. F. (1993). Using psychosocial and cognitive behavioral theories to promote self understanding: A beginning framework. *Journal of Counseling and Human Service Professions,* 7(1), 8–18.

Combines Erikson and reaction patterns and discusses implications for human service work.

Weinstein, G. (1981). Self science education. In J. Fried (Ed.), *New directions for student services: Education for student development* (pp. 73–78). San Francisco: Jossey-Bass.

An easy-to-follow model for interrupting reaction patterns that may be troubling you.

ON SYSTEMS CHANGE

Homan, M. (1994). *Promoting community change.* Pacific Grove, CA: Brooks/Cole.

Excellent, well-written text on community organizing with interesting thoughts on the change process.

Watzlawick, P., Weaklund, J. H., & Fisch, R. (1974). *Change: Principles of problem formation and problem resolution.* New York: Norton.

Somewhat theoretical but very interesting. A groundbreaking book on why change can be so stubborn and what to do about it.

References

Corey, G., & Corey, M. (1997). *I never knew I had a choice* (6th ed.). Pacific Grove, CA: Brooks/Cole.

Lacoursiere, R. (1980). *The life cycle of groups: Group developmental stage theory.* New York: Human Sciences Press.

Lewin, K. (1954). Behavior and development as a function of the total situation. In L. Charmichael (Ed.), *Manual of child psychology* (2nd ed.). New York: Wiley.

McClam, T., & Woodside, M. (1994). *Problem solving in the helping professions.* Pacific Grove, CA: Brooks/Cole.

Sweitzer, H. F. (1989). The BPE framework: A tool for analysis and interdisciplinary integration in human service education. *Human Service Education,* 9(1), 11–19.

Watzlawick, P., Weaklund, J. H., & Fisch, R. (1974). *Change: Principles of problem formation and problem resolution.* New York: Norton.

SECTION FOUR

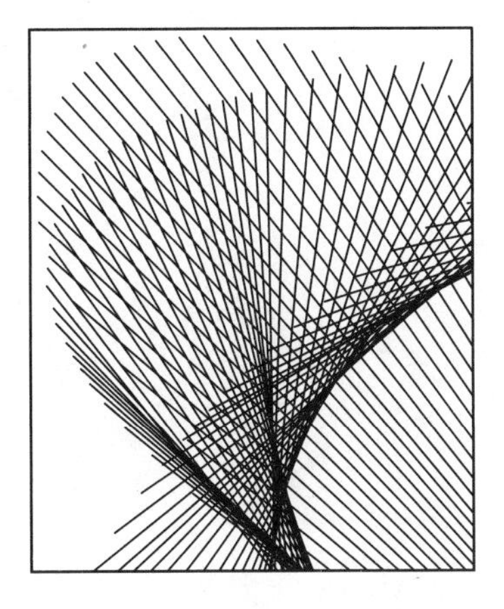

GOING THE DISTANCE

You have come quite a distance in your internship. You have dealt with the concerns of beginning, you have launched and nurtured a number of relationships, and you have learned how to identify and cope with problems. The next several weeks ought to be a time of great productivity and enjoyment, although they will undoubtedly have their challenges. Chapter 10 examines in more detail the joys and challenges of the Competence Stage in your internship. But even as you are enjoying this time in your internship, thoughts of ending will begin to creep into your consciousness. This is yet another critical time in the internship. Ending well is not easy, but it can be done. Chapter 11 will help you meet the challenges of the Culmination Stage.

CHAPTER 10

Riding High: The Competence Stage

Things that didn't seem to come together have finally come together.
STUDENT JOURNAL ENTRY

The price of greatness is responsibility.
WINSTON CHURCHILL

There comes a time in almost every internship when interns tell us that they are enjoying their internships in a whole new way. The initial anxieties have subsided, problems big and small have been resolved, and considerable confidence has been gained. If this sounds like what you are experiencing, then you have entered the next stage in internship, the Competence stage. Like any stage, this one has its challenges, but for many interns and supervisors alike, it is the most exciting time in the field experience.

We use the concept of *competence* to refer to the intern's overall productivity and achievement, including interpersonal and intellectual skills (Chickering, 1969). The intern's concerns during this stage focus on developing a sense of competence and taking charge of the internship to ensure a quality experience. Striving for competence often means being all that you can be in your role as intern and doing your very best. You may recall the discussion of competence in Chapter 2 and Erikson's (1963) position that a sense of competence includes feeling like a capable person even when you stumble or fail. Most likely, you have experienced times during your internship when you were disappointed by your performance in some way. According to Erikson, if you have a strong sense of competence, moments like these will not shake the basic confidence you have in your ability to succeed and move ahead.

You may be realizing by now that you have grown not only in terms of your knowledge and skills but in personal ways as well. The transformation that is occurring may seem slow and subtle, so much so that you didn't notice until someone commented on it. Or it may seem sudden and dramatic, leaving you to wonder what is happening. Either way, you are changing. You have successfully faced new situations, including ones that seemed overwhelming at the time; you have accepted greater and greater responsibilities, developed your competencies, and enhanced faith in yourself; and you have also learned to develop new ways of thinking about situations and the people with whom you work, including yourself. None of these changes has come easily, but they add up to a transformation that lets you see more, do more with what you see, and be more in charge of your internship. In short, you are empowering yourself, and it shows every day in your work.

Students arrive at this stage at different times in their internships. Some of your peers may not be here yet, while others may have been here for some time. What is important is not how fast you reach this juncture, but rather having confidence in yourself that you will get here in due time. As you join your peers in reaching the competence stage, keep in mind that you have different styles of expressing your emotions (see Chapter 8), and while you may experience this stage as a subtle, yet profound, shift in feelings, others may experience it as a positively giddy swing in emotions. In this chapter, we review some of the pleasures you can expect from this stage and prepare you to face the challenges that lay ahead.

ENJOYING THE RIDE

This is the time in field placement when your experience most closely matches the images you had of what an internship was all about. You finally are doing what you set out to do. You have greater responsibilities each week (and may even be wanting more!) as you strive to reach your goals, and most of your energy these days is devoted to fine-tuning your skills so that they become second nature and you can generalize them to everyday situations.

Chances are that you are also seeing your internship through new and different lenses. This is often a time when interns are able to differentiate more clearly between what is important to their work and what is not (Dreyfus & Dreyfus, 1980). Subtle dynamics that escaped you before are probably visible now. Remember when you were concerned about acceptance issues? Even though you spent quite a bit of time trying to find your way around the organization while you had these issues, the politics were virtually invisible because your primary concerns were elsewhere. Now, however, you are more savvy and in tune with the ways of the organization, and consequently may feel office politics when they are affecting your work because your primary concerns are now on quality and competence in your work.

Your emotional landscape is changing as well. Many interns feel an enhanced sense of mastery as their skills become sharper and their sense of autonomy requires less constant supervision and direction. There is also a tremendous sense of confidence associated with this stage. The anxiety, awkwardness, and trepidation of earlier stages have gradually given way to a period of calm and inner strength. Interns often tell us that things have settled down at the site. However, life at the agency is still hurtling along at the same pace. It is not the world around you that has changed and become more peaceful; rather, it is you who have developed an inner sense of calm from which you will derive strength as you move through this intensely productive period in your internship.

Redefining Supervisory Relationships

The importance of the supervisory relationship cannot be understated: It is the aspect of the internship that our students have consistently reported over the years as most critical to success in the field. We are sometimes struck by the quality of the relationships our students develop with their supervisors. They not only are supportive and encouraging connections, but they often become mutually satisfying relationships that last long past graduation day.

When we ask our interns to reflect on the changes that have occurred in their field placements, they immediately identify changes in relationships with their supervisors. They smile as they tell us that they no longer need to check in with their supervisors to do their work. Rather, supervision occurs at weekly, structured, individual/group meetings. This shift in frequency of and dependence on the supervisory meetings can be most affirming for emerging professionals like yourselves.

Working with staff and supervisors becomes easier during this stage: Communication becomes more comfortable and open, issues can be approached without concern about rejection or conflict, genuine teamwork often develops, and supervision can be a source of insight and feedback for personal as well as professional growth. Often, a change in the openness of the supervisory relationship becomes evident at this time as well. Most of you would not have considered challenging your supervisors on theoretical grounds earlier during the placement. However, a combination of confidence and a sense of emerging equality may make a big difference in your willingness to engage the supervisors in academic discussions as well as disagreements (Lamb, Baker, Jennings, & Yarris, 1982).

There has also been a shift in your primary focus. Chances are you are talking less about yourself during supervision and in seminar class and more about the clients' needs and the helping relationships you have developed with them. This is just one more shift that tells you important changes are taking place (Tryon, 1996). One explanation for such shifts is that your supervisory needs are also changing (Hersey & Blanchard, 1982; Tryon, 1996). Your former dependence on your supervisor for direction and support has given way to working more autonomously and developing more equal relationships with your supervisors (see Chapter 6 for a review of supervision issues).

Some site supervisors develop more than a supervisory relationship with their interns. They develop what is traditionally called a *mentoring relationship*. The mentoring relationship differs from supervision because it is interpersonal in nature and is one in which the supervisor takes an interest in the intern's professional development and career advancement (P. Collins, 1993). Many studies have been conducted to understand what happens when this type of relationship develops. It can be very empowering, leaving both the mentor and the mentee better professionals for it. It may interest you to know that the mentoring relationship is often considered the most important aspect of graduate education (Mozes-Zirkes, 1993). At the undergraduate level, we have found that the mentoring relationship, when it develops, is a most instructive and influential factor in the quality of the internship experience.

The best mentoring relationships occur spontaneously between the supervisors and the interns. The mentoring supervisor makes sure that the intern becomes part of the organization very quickly, is given highly visible tasks, and is introduced to the profession through such resources as networking, luncheons, and conferences. Although many supervisors do make sure that all of these bases are covered for the intern, the mentoring supervisor personally invests considerable time and many resources in coaching the intern for success (Tentoni, 1995).

Of course, this relationship, like others, takes time to develop (P. Collins, 1993). It moves from a relationship of positive role modeling, when much of the learning occurs through interactions, observations, and comparisons with the supervisor, through a time when you have grown to really like each other, and then to a point where you might be right now: of valuing the relationship and recognizing how mutually rewarding it is for you and your supervisor. When this happens, the intern tends to become increasingly self-assured, competent, and autonomous, no longer needing the mentor in the same ways for guidance or support, and eventually becoming more independent of the supervisory relationship. In the best of mentoring relationships, a more equal relationship emerges over time. That becomes evident when the supervisor accepts you as a colleague, and you accept the supervisor on equal footing (P. Collins, 1993).

Redefining Your Identity

A second area of significant change is a most visible one: your career status. A transformation is taking place, and it is a change in your status from that of a student to that of an aspiring professional. You think differently, you act differently, and you feel differently. How did these changes happen? You may not be aware of this, but how you think about yourself is redefined by your experiences in the internship and the situations you face every day at the field site. You no longer refer to yourself as "just an intern" or "the intern"; rather you recognize yourself more and more as staff and as a member of the profession. One intern reported, "I am fitting in around the office. Everyone is helpful and giving me tips on how to do things. They even ask me what I would do in certain situations. I feel they respect what I have to say." You have a greater sense of self-awareness and self-respect. You are most likely learning that you can trust your knowledge and skills, and you are developing a good sense of your strengths and limitations. Your work might suddenly be taking on new meaning because you realize that you are making a difference in the lives of others. Your professional identity is beginning to emerge.

There are two hallmarks of this shift in identity that are discussed with frequency by our students: a commitment to quality work and a commitment to personal integrity. Not so long ago, "good enough" may have been all you expected of yourself in your work; just getting the work done was your goal. That was especially so if you were dealing with a multitude of responsibilities in addition to your internship. Now, though, the "good enough standard" that prevailed at the beginning of the internship gives way to a standard of excellence as you enter the Competence stage. You become more demanding of yourself regardless of the amount of responsibilities you have. This expectation of excellence extends not only to all aspects of your internship, but to those aspects of your personal life that are directly connected to the field experience.

We do not think an internship necessarily changes your basic core of values; however, it certainly does push you to clarify them professionally and personally. When this happens and you are accomplishing a great deal, you may feel so proud of your work that you want to keep on doing it and doing it better. A word of caution here. Many students have learned from experience not to confuse the need to do work well (excellence) with the need to do work perfectly (perfection). Though both may guarantee success, one guarantees headaches as well. If perfection is your only or preferred way of doing things, you are bound to have problems.

The second hallmark of a changing identity is a more conscious commitment to a sense of personal integrity. Changes here are reflected in the ways that you think (flexibility and adaptability), what you value, and how your beliefs and values affect your behavior (Chickering, 1969). Although you may feel that you have matured substantially since you began your internship (much to your relief and that of your family!), changes in your basic level of integrity tend to be gradual and do not affect the values that are important in your life (Chickering, 1969). Developing integrity is a lifelong process that begins during the college years. This is not a goal of your internship, but rather a gift that emerges from the maturing process. As you grow in integrity, your values become genuinely reflected in how you act. In other words, your behaviors truly represent your values. For this harmony or congruence to occur, it means letting go of your literal beliefs in the external rules you live by and reconstructing a new value structure for living. This is no small feat, and in fact, you already are working on it. By doing so, you are "humanizing values," which is essential for ongoing maturity. Though necessary, letting go of one's literal beliefs can be quite perplexing when in the throes of a life transition, such as the internship is for many students.

Because anxiety and tension can run very high at these times, students develop an internal structure to make sense of their world. This new structure that emerges is one that represents the essence of you. It is not your parents', your school's, or your friends' ways of thinking, but yours and yours alone. You own it (emotionally speaking); it is a part of you and it reflects your values and beliefs. The closer you are to adopting this value structure openly and without façade, the closer you are to living with a sense of personal integrity (Chickering, 1969). This is a lifelong task, and your journey has only begun.

BRACING FOR THE BUMPS

Although the competence stage is recognized primarily for its good feelings and intrinsic rewards, it has its challenges like all other stages in the internship. For some interns, the highs of this stage allow the challenges to be put into perspective and managed rather easily; for others, the challenges seem to threaten the good feelings that characterize this time in the placement. These challenges cannot compare to the ones you faced in the past; those felt like walls that stopped you from moving forward, whereas these are more like bumps along the way. A word of caution is needed here, though. These bumps in the road can throw you off course if you do not know what to expect or how to manage them at the time. We have found three potential bumpy areas that you need to pay attention to: (a) the tendency to focus on only the good feelings and avoid the challenges that are an inevitable part of your placement, (b) the tendency to not enjoy the success you are achieving, and (c) the predictable issues inherent in a transition to this highly productive time in your internship.

Freezing the Moment

Many interns fall prey to the misconception that the days of difficulties or problems are behind them. Some say they expect more difficulties in their internships, but when the difficulties arise, they ignore or otherwise do not acknowledge them. Other interns just want to focus on the good feelings of having resolved issues. And still others fall into the trap of believing that real professionals (like really good interns) should be beyond having difficulties.

Well, you are not beyond them—no one is. Your skills, your relationships with clients, supervisors, peers, and co-workers, and your understanding of the field site are all evolving, and you could continue to experience new difficulties in all these areas. The good news is that you are better equipped than ever to meet these challenges; in fact, your ability to handle difficulties can be a source of ongoing pride. The difficulties you face at this time may actually feel different now and not be so overwhelming or anxiety provoking. However, they still need to be managed or the difficulties will become problems.

Experiencing Success

Some interns become challenged at this point in their internships when they realize that they appear successful to others but they do not *feel* successful about their accomplishments. Such a situation can become problematic because your experience of the internship is based on what you feel, not on the perceptions of others. So feeling good about all aspects of the internship is very important to your sense of success. Being able to feel your success, often described as a sense of fulfillment, is what makes the success genuine; anything less can compromise it (King, 1988).

What is particularly interesting is that this sense of inner success, which reflects your success as a person, is believed to render all of your achievements worthy (Huber, 1971; King, 1988). Not only do you need to achieve the goals of your internship (outer success), but you also need to pay attention to yourself as a person (inner success) if you want to reap such psychological rewards as the internship being "the best experience of my life" and "a most fantastic experience." In the section that follows, we discuss three sources that nurture the emotional experience of success (King, 1988). As you read about them, think of your own field placement and the ways it provides you with what you need to feel the success you have earned.

The first source of fulfillment in success concerns the "outer," or social, dimension of success (Huber, 1971) and involves the work you do in your internship every day. Whatever your work might be, it must be considered *worthwhile work* by you and your supervisor. The work itself needs to include productive and responsible activities that have personal meaning, and it must allow you to accomplish clear goals.

There are five aspects of the work that allow these feelings to develop. First, the goals of the learning contract need to reflect *accomplishments* that are purposeful, constructive, and challenging, which can be a source of pride for you. The work itself needs to provide you with opportunities to develop your skills and apply concepts in real life situations. For your part, you need to take an active role in creating your work

FIGURE 10.1
Sources of Fulfillment in Success for Interns

Doing Worthwhile Work	Developing Responsible Relationships	Defining Success for Yourself
Accomplishments	Supervisors	Conscious Choices
Acknowledgment	Peers	Active Involvement
Self-Determination	Staff	Self-Determined Goals
Self-Actualization		
Intrinsic Rewards		

and your achievements. Values play an important role in the sense of accomplishment you feel about your work and your association with the agency. Your values and those of the field site need to be compatible for you to feel like a part of the organization. In addition, both your work and that of the agency need to reflect socially responsible goals to have credibility in the community. Interning at a site that is a known abuser of the environment or has a reputation for questionable practices with clients would certainly call into question—by you and by others—the quality of your success.

Next, it is important that you be recognized and respected by your supervisor; such acknowledgment is especially significant when your contributions are above and beyond what is expected of an intern. The third important aspect of the work has to do with the potential for *self-determination* (autonomy); for example, it is important that you have the freedom to create and carry out tasks. And the fourth aspect has to do with the potential for *self-actualization* in your work; for example, there need to be opportunities for creative expression and personal growth in the work you do. Above all, the work has to be *intrinsically rewarding*, such as when you make a difference in the lives of clients and communities. Then, regardless of the demands or stresses, the sense of intrinsic worth will allow you to derive pleasure from what you do.

The second source of fulfillment in success concerns the "inner," or personal, dimension of success (Huber, 1971) and is derived from the personal relationships that are integral to your experience of the internship. For your relationships to be a source of feelings of inner success, they need to be genuine, cooperative, and mutually satisfying for all involved; in other words, they need to be *responsible relationships*. Real commitment on your part and on the part of staff, co-workers, peers, and supervisors is needed to make that happen. The staff must be receptive to an intern coming on board. The supervisor needs to establish an effective supervisory relationship with you and include you in work groups/teams, staff meetings, and social functions (when appropriate), and you should be willing to engage both the supervisor and the staff in responsible supervisory and collegial relationships.

The third source of feeling the success you attain is being able to *define success for yourself*, which means you are doing what you want to do. This choice must be yours and yours alone, and it must be a conscious choice made free of the influences of other people. When it comes to your internship, it means having been actively involved in the selection of your field site and in developing the goals of the learning contract. If the campus supervisor determined your site and the site supervisor determined your learning contract, and you were not actively involved in an informed way in these processes, the likelihood of feeling successful about your internship may have been compromised. However, even if you went into the internship not consciously aware of what you really wanted, those overseeing your actual placement may have been able to assess your interests and goals well and place you with a supervisor who provides considerable support for personal growth and awareness, while at the same time entrusting you with worthwhile work. You will need to do some soul searching about this, and now might be a good time. Think about what you really wanted to achieve in this internship, rather than what others think or wanted for you, to what extent your placement site is a good match for your personal goals, and to what extent the choice of internship was made freely by you.

So, what do you do if you are not feeling your success? Actually, there are several things you can do. First, you can spend some time determining whether or not the design of your internship allows you to experience the steps of processing and organizing experiences so that the learning cycle described in Chapter 2 takes place (Kolb, 1984). How does the design of your internship measure up to that framework? Next, you can spend some time examining the relationships with your supervisors and others who are important to your field experience. Are they disappointing relationships in any way? If so, why do you think that is? If not, what makes them so good? Finally, you can review the sources that nurture your feelings of success and, using the Steps to Create Change from Chapter 9, identify the areas in your internship that could be strengthened.

Encountering Transition Issues

There are several issues that tend to surface as you enter this period of highly productive work. They all seem to be part of a natural transition to a more realistic way of living with the internship once the crises in growth are over and the climb has ended. If these issues are not recognized and understood for what they are, they can compromise your feelings of competence and the joys that come with this stage in the internship. A discussion follows of three of these issues: (a) the leveling off effect, (b) the need for a more balanced life, and (c) the inevitable "crunch."

The *leveling off effect* is experienced as a subtle change in the pace and intensity of your ascent on the ladder of responsibilities. Students often liken it to that point in an air flight when the plane is no longer ascending and begins to level off. The timing of this change in the internship is determined either by the calendar or by a saturation in work load. When this happens, you will experience a cruising effect: The intensity of increasing responsibilities decreases and you no longer feel the constant push to greater heights. Eventually, you will stop taking on new responsibilities and settle into a rhythm of working that is more realistic and predictable in nature.

For many students, leveling off means that the goals of the internship are being reached, and it is time to relax somewhat and enjoy the ride. For some, though, it is difficult to enjoy the ride because of the mixed feelings that come with reaching the final plateau of a journey. For a brief moment, you might wonder whether this is all there is to the internship and speculate that there are few surprises left in the work and little left to learn. Be assured that any emotional letdown you experience in reaching the apex of your journey is only momentary and will pass. You will adjust to these changes and get on with the challenges and personal rewards that come with being a productive member of the staff.

You are probably sensing by now that your total immersion into the role of being an intern is beginning to change (Lamb et al., 1982). If you are like many interns, you may suddenly realize that you had a life before the internship—a social life, a family life, a private life—and you find yourself needing to reclaim that life. Perhaps that awareness will come to you on your way home from the field site, when you are no longer so preoccupied and are thinking about aspects of your life that were left behind in all the excitement of the internship. Or perhaps it will happen during a morning break as your mind wanders to more personal needs—for relationships, for family time, for time alone. "I can feel a tug going on between family demands, work, and my academic requirements," reported one intern. "The holidays can also complicate my situation to the point where I am beginning to feel both pressure and stress." Interns usually find a need to get on with living at this point in their internship. In our experience, these changes signal a healthy channeling of energies toward re-creating a more *balanced way of living* with your professional life. Though your internship still remains very important, it may no longer be the driving force of your energies.

Up to this point in your internship, you have had to struggle with the ups and downs of becoming part of an organization, learning skills, developing competencies, resolving differences between real and ideal expectations, and contending with changing perspectives and life's big questions. Just when you thought it was clear sailing ahead . . . *wham!* You've run into yet another wall, and just about everyone hits this one. This wall is different, though. You know you can manage it; you just don't know whether you can muster the energy you need to manage it. You feel some anger about the amount of work you have and frustration over the amount of time you do not have. Something subtle is happening as well. Your perspective seems to be changing again. You have little tolerance for trivialities, and you find yourself feeling indifferent about some of your assignments and responsibilities, especially if they are not directly related to the internship or do not measure up to the importance of the work you are doing in the field. This is not a crisis in growth you are having; this is a crisis in the management of your time and work load. It takes on added importance because it in fact is a threat to your schedule and to the pace of your internship.

You are encountering what we refer to endearingly as *the crunch.* It is an early warning signal to let you know that the end of the internship is fast approaching. It really is no different from the crunches you have faced at the end of every semester (usually at the end of the midsemester slump, remember?) when papers, exams, and projects all become due at the same time. However, this time, much more is at stake academically, financially, and professionally. You also have the greatest number of responsibilities you have had in your academic career. You begin questioning the merit of each assignment, challenging all requests for more work, and trying to not crash into the wall of professional ineptitude. You are beginning to feel overwhelmed and frozen in time . . .

too many responsibilities . . .

too many deadlines . . .

too many details . . .

too few resources . . .

too little social life . . .

too little family life . . .

too high standards . . .

Oooops! You're beginning to slide . . .

If any or all of the preceding items best describe your internship at this time, you are not alone! Most of your peers are going through the same crunch in their internships. It is time to rethink your situation and regroup yourself, emotionally and physically, for the last mile of the journey.

You know yourself best in times like this. What works for you? What allows you to manage time and work load crunches? Some people actually thrive when they are up against a wall of deadlines, while others crumble under the same circumstances. Perhaps one of the following suggestions will work for you, as they have for many of our students.

Take time for yourself A day will do; an afternoon or evening could do just as well. The important thing here is to stop. Just stop everything. Take a holiday from the assignments, the responsibilities, and the time lines you are facing. Of course, it is not just the internship that is overwhelming you. Many of you have many more responsibilities beyond your internship, and you may be overloaded for what you can effectively handle at this time. However, the internship is still the new family on the block, and its work load is most demanding right now. It is very important that you do something that you enjoy. Sing. Dance. Shop. Bike. Hike. Swim. Draw. Paint. Read. Run. Do something, anything, whatever it takes to help you let go and relax. Try to do this once every week.

Prioritize and schedule You must look at the calendar and make some decisions. It is time to prioritize your tasks because things simply must get done. What you will need to determine is what you can do in the time you have and what can get done only if you make big changes in your schedule. Here is an exercise that our students find helpful in accomplishing this task. First, list all the tasks that you have to complete. Then, next to them, indicate the time frame within which they can be finished, not necessarily must be finished. For example, it you have a final culminating statement to write for your seminar class, it obviously cannot be completed before the internship ends. However, you can work on much of it before the end of your placement. Then, when the internship is over, there is relatively little content to add. The time frame for this assignment actually begins on the date you start this exercise and ends on the due date. By listing tasks, assignments, and responsibilities in a similar manner, you will have a more informed framework for prioritizing tasks and assigning calendar dates to work on them.

What is obvious is that something has to change to put you back on track and in charge of your internship. You already had to deal with difficult challenges when you confronted the barriers to your internship (see Chapter 9). Using the Eight Steps to Creating Change model, try to clarify the biggest hurdles facing you in this management crunch and identify ways you can prioritize them to bring the crisis to an end.

ETHICAL ISSUES AND CONCERNS

When the focus of your concerns is on competence, you can also come across issues that make you unsure about what should be done. Often, interns tell us that they encounter situations in which they have a deep sense that something is not right. However, they may have neither the language nor the knowledge to identify an issue for what it is and deal with it in an informed way. Instead, they tend to become confused by complex issues, especially those with ethical and legal dimensions, and are at a loss as to what to do about them because they often have little background or training in such matters. These situations are complicated, ambiguous, and often lack clear answers; they are challenging for even the seasoned professional. Although a full discussion of these issues is not possible here, we will cover them briefly and refer you to resources should you wish to pursue them on your own.

The following terms, which are used when talking about ethical, moral, and legal situations, are defined to make it easier for everyone to participate in a discussion. An *issue* refers to a point that is in question or in dispute, and *professional* refers to matters that pertain to an occupation. *Professional issues,* then, refer to some aspect of how one goes about doing one's work that has become a matter of debate among others. *Standards,* when used generically, refer to guidelines or codes that govern the behavior of members of a given profession. *Ethical* suggests that someone is acting in accordance with professional standards, codes, guidelines, or policies, and *unethical* suggests that they are not. *Legal* suggests that someone is acting in accordance with the law, and *illegal* suggests that they are not.

In addition to these terms, which are related directly to the responsibilities of a profession, the terms values, moral, and ethics are very important to any discussion on decision making about professional, moral, ethical, and legal issues. *Values* refer to what is intrinsically good, useful, and desirable; *moral* refers to what is right or wrong conduct in its own right, based on broad mores such as religious principles; and *ethics* refers to the moral principles or rules of conduct of a particular group, such as human service ethics, criminal justice ethics, mental health ethics, education ethics, and so on (G. Corey, Corey, & Callanan, 1998; Pollock, 1998).

In addition to a common language, access to common resources facilitates discussion and decisions about ethical issues. For one thing, although your understanding of the professional issues in human services work may have just begun, you are still responsible for acting in accordance with the values and standards of the profession. These values and standards are embodied in documents variously referred to as guidelines, codes, policies, regulations, and laws. It is precisely these documents, which are reviewed when professional situations arise, that do not lend themselves to clear responses, interpretations, or resolutions.

A word of caution is needed here. We have found that when students review the professional standards, codes, or guidelines that govern their dimension of human service work, some of them (especially those who are practitioners) become angry with what they consider to be the esoteric nature of the statements found in the documents. The statements they read often seem unrealistic and not at all sensitive to the demands of their work. Students tend to question the sources of the documents and sometimes ignore them as guides for work. Many of these professional documents are designed to be either *minimal* (mandatory) or *ideal* (aspirational) standards. A reader might not be aware of whether the document in question uses the minimal or ideal approach; possibly, it mixes the two. As you examine all such documents, be sure that you know their intent. It makes a big difference if you read as minimally expected behaviors those which are ideally intended, or vice versa.

The *Ethical Standards of Human Service Professionals* (National Organization for Human Service Education, 1996) is a generic set of broad guidelines for the many disciplines across the human service profession. We hope that you take the time to actually read through this document before moving on in this chapter. The standards will clarify responsible ways to carry out the work of the profession. They will also lend considerable insight to the following section, where we identify many of the issues that could potentially become problematic for the intern and the practitioner. However, you may be more clearly aligned in your work with a specific discipline within human services, such as social work, counseling, rehabilitation, or criminal justice. If so, your responsibility is to be aware of the documents specific to that area as well. Your site supervisor or supervising professor can provide copies if you do not yet have them. It is also a good idea to have access to the names and addresses of all the professional organizations to whose codes you may be held accountable. Those most directly related to human services are noted at the end of the chapter.

There are a number of perspectives that you can take when considering an issue that will prove useful to your work. For example, let's take the issue of having sex with a client. Common sense tells you that this behavior is inappropriate. However, many professionals have been accused, have been caught, or have acknowledged doing just that. Furthermore, the issue is not something that everyone can agree on. Some members of the profession see little wrong with such behavior under certain circumstances; some find it reprehensible under all circumstances; and others have opinions that are somewhere in-between. However, when this issue is considered in terms of the four perspectives of legal, ethical, moral, and professional, it is generally found to be unacceptable on all accounts: illegal, unethical, immoral, and unprofessional (G. Corey et al., 1998). There is little disagreement on this issue when considered in this manner. We agree with G. Corey et al. (1998) that the standard of "client's best interest" is one that best allows the practitioner to maintain a clear ethical position on issues. When in doubt, always ask yourself: Is what I am doing, or what I am seeing or hearing about, in the best interest of the client?

In the following section, we consider some issues that may arise as you observe others in the workplace. Then, we examine some of the more common dilemmas that you may face in your own work in the internship. Finally, we offer a model for making decisions when confronted by these issues.

Questioning Conduct

Once their concerns focus on developing competence, it is quite common for interns to pay attention to the professional conduct of their co-workers as well as their own. In particular, interns tend to pay a lot of attention to the ways in which the staff members go about their work, deal with clients, and conduct themselves with colleagues. You might find yourself beginning to look at the staff and your supervisors differently now. They are not just professionals with roles; they are professionals with moral, ethical, legal, and professional responsibilities to the profession, the clients, the organization, and you. You are now tuning into the subtleties of their behaviors and becoming aware of possible improprieties in how those responsibilities are being met.

The improprieties that you may observe did not begin when you first noticed them. In all likelihood, it is you who have changed: You can now see what has been there all along. The behaviors and attitudes were just not within your awareness. There are a number of possible reasons for this change. First, staff tend not to disclose questionable or surreptitious aspects of themselves so readily with interns or new employees, but rather tend to act as they are expected to act in their roles (Kanter, 1977). Consequently, they tend to shield questionable behaviors and attitudes from interns until they get to know them better. Another explanation is that you have been so busy with your increasing responsibilities that you have not had time to notice these behaviors or even think that there was something to notice. A third possibility is that you held stereotypes that needed to change before you were willing or able to see situations for what they were.

There are a number of situations that lend themselves, conditionally or not, to questionable behaviors. Some situations are so obvious that mentioning them seems absurd. However, they do need to be mentioned because aspiring professionals, like seasoned professionals, are people, too, and they have personal frailties that compromise their ethical standards at times. We both have known and worked with individuals who committed improprieties that neither we nor they ever would have expected. The following list (in part from Royse, Dhooper, & Rompf, 1996) mentions some of the more common of these improprieties:

- being sexually intimate with clients or supervisors
- libelous or slanderous actions against clients
- threatening or assaultive behaviors against clients or co-workers
- misrepresenting one's status or qualifications
- abandoning a client in need of services
- failing to warn and protect appropriate parties in the case of a violent client
- failing to use reasonable precautions with clients dangerous to themselves
- being dishonest or fraudulent in your actions

Regardless of the reason, you are bumping up against issues now that did not concern you in the past. And they can become complicated enough to be potential pitfalls for any aspiring or seasoned professional. When the issues are of a legal or ethical nature, the situation can become precarious and needs to be managed with reason and sensitivity. What is at stake is at least someone's reputation and at most someone's career. When the stakes are that high, they are high for you, too, because your opportunities for employment could be compromised if your concerns are not justified and handled professionally.

Grappling With Dilemmas

One of the most wrenching aspects of working in the field of human services is dealing with a dilemma that involves the welfare of another individual, family, or community. A dilemma—be it ethical, legal, moral, or professional—refers to a struggle that occurs among alternative courses of action that might resolve a situation. To complicate things, the courses of action tend to be correct in their own right, but they conflict with each other. A dilemma, then, is a situation that you can find yourself in whereby you have more than one justified course of action to take on an issue.

Our interns tell us that once they develop an understanding of the issues and become comfortable with the language used to talk about them, they begin to see the issues in their daily work. It just so happens that you develop this awareness at about the same time that you may be moving into a peerlike relationship with your supervisor. Consequently, you are much more apt to talk about incidents, behaviors, and concerns at this point in your field experience than you were a few months ago. Once you are aware that a situation in fact is an issue, the next hurdle is to recognize a dilemma when you see one, which is no easy feat. Our experience tells us that recognizing dilemmas as such is quite challenging for both undergraduate and graduate students, as well as for professionals, regardless of how long they have been in the field. Often, students cannot readily name ethical issues when they see them, and they do not necessarily see them in a given situation. Academic programs in human services education sometimes require a course in professional issues, but more often will offer it as an elective course or cover the content in other coursework (M. DiGiovanni, personal communication, October 1997). Feeling particularly unprepared for these challenges makes sense if you have not been prepared academically for them.

There are three basic types of dilemmas: (a) those that result from your own decisions, behaviors, or attitudes; (b) those that result from another person's decisions, behaviors, or attitudes and directly affect you; (c) those that you observe from a distance, such as what you have seen or heard, but which do not directly affect you. An example of the first type is considering whether to engage in a dual relationship with a client, such as knowingly working with a client whose sister you have dated. An example of the second type of dilemma is working with a client who, unbeknownst to you, is your cousin's intimate partner. An example of the third type of dilemma is observing or hearing about a co-worker dating a relative of his or her client. These three types of dilemmas can be seen in all four categories of the following issues.

GRASPING THE ISSUES

Beginning in the late 1980s, literature has proliferated in response to a growing need for workers to know about rights and responsibilities in human service work so as to better serve the consumers of services. Although a detailed discussion of these issues is beyond the scope of this book, a very helpful way of organizing the most commonly recognized issues is to think of them in terms of distinct aspects of your work on-site (Chiaferi & Griffin, 1997). We use this approach to organize our discussion of the most common issues facing interns in the field of human services. It has been adapted to meet the needs of this discussion and incorporates issues across the spectrum of human service work from multiple sources (American Psychological Association, 1995; Baird, 1996; Chiaferi & Griffin, 1997; D. Collins, Thomlison, & Grinnell, 1992; G. Corey et al., 1998; Goldstein, 1990; Gordon & McBride, 1996; Martin, 1991; Schultz, 1992; Wilson, 1981).

In each of the four categories of issues that follow, we identify a situation that practitioners may deal with in their careers. Some of the issues actually overlap due to the nature of the work in different dimensions of human services. We have listed them in detail so you can see in print the vastness of the responsibilities of your work. You are expected neither to remember them nor know them, though in time and through practice they will become second nature to you. At the very least, though, you will know where to go to find out more about them. For the most useful and comprehensive treatment of the subject, we refer you to the seminal text *Issues and Ethics in the Helping Professions* (5th ed.) (G. Corey et al., 1998). Table 10.1 shows some of the more common issues that you may encounter, grouped by category.

Professional practice issues concern how people approach their professions. An example of a dilemma that falls into this category is your having information about a staff member who is demonstrating insensitivity to the culture of a client's family, is misrepresenting qualifications to work with such families, and is now being promoted to supervisor of your work group at the site. It happens that you and the client share a similar cultural identity. It also happens that you and this staff member have had difficulties working together in the past. You are wondering whether you should disclose your concerns to your site or faculty supervisors, talk directly with the staff member, or say nothing.

Generic human service issues deal with how people approach the work of their professions. An example of a dilemma in this category is your being out to dinner and overhearing the conversation of an agency worker who does not know you, but whom you recognize. The conversation is about a client, and the worker identifies enough data that you recognize the case from a staff meeting earlier that week. You know from your orientation period that the agency has a policy that workers are not supposed to talk about cases outside of the office, or in the very least, they must not disclose any identifying information about the case. The person with whom the worker is speaking is your child's teacher. Neither of them saw you, and you are not sure what to do. You realize that something inappropriate has occurred, but you are also hoping to be hired at the end of your internship and are hesitant to pursue the issue for fear of making the wrong move.

Service intervention issues deal with working directly with clients. An example of a dilemma that would fall into this category is finding out at an Alcoholics Anonymous meeting, which you attend for personal reasons, that your supervisor's client is planning to leave the country in the next couple of weeks. Of particular concern is the fact that a friend told you that she overheard the client threatening to hurt a former girlfriend. You had been told that what goes on at Alcoholics Anonymous meetings is confidential, and you know how strongly confidentiality is valued in your future profession.

TABLE 10.1
Ethical Issues Facing Interns

Professional Practice Issues
- competence in doing the work
- frequency and focus of supervision
- consultation
- education
- diversity awareness
- grievance issues
- limitations in the scope of practice
- credentialing/license standards and requirements
- advertising for services
- dressing for the role
- relationships with supervisors and staff
- managing the risks of physical danger and legal liabilities
- disclosure of status as student
- personal disclosures
- criminal activities
- political influences/corruption
- subversion of service system

Generic Human Service Issues
- dual status/sexual relationships with clients
- record keeping
- informed consent
- privileged information
- right to privacy; confidentiality
- values of benevolence, autonomy, nonmaleficence, and justice
- exceptions to confidentiality, including abuse/neglect cases
- dangerous-client cases (self and others)
- third-party payer requests
- court orders for release of information
- working with minors
- client requests for release of information
- duty to warn and protect
- the integrity of clients

Service Intervention Issues
- clinical issues (transference and countertransference)
- limitations on scope of responsibilities
- client's right to self-determination
- management of referrals
- size and nature of caseload
- termination

You wonder how to uphold your responsibilities to all the parties involved (i.e., the client, the girlfriend, A. A., and your profession), many of which seem to be in conflict, and how to determine if some take priority over others.

Internship issues deal with your internship and you. The potential for liability here is very real. Interns may deal with all the issues identified in the previous three categories and are held to the same standards as employees. In addition, there are a number of issues that are specific to experiential education and internships in particular. For example, as an intern, you have the multiple roles of being a student, a consumer of services by virtue of your need for supervision, and a care provider to others. Accountability is demanded in all three areas.

An example of a dilemma in this category is your becoming aware of unethical and possibly illegal practices at a certain field site that you want for your internship and

TABLE 10.1
Ethical Issues Facing Interns *(continued)*

- abandonment by therapist
- release of information
- sharing information with colleagues
- emergency response during nonworking hours
- differences in legal and ethical practices
- individual vs. group vs. marriage and family interventions

Internship Issues

- rights to supervision
- responsibility to confront situations in which educational instruction is of poor scholarship and nonobjective
- disclosure of risk factors to all potentially affected parties (to site supervisor about intern; to intern about site supervisor)
- behaving consistently with community standards and expectations
- awareness of risk status of agency
- active involvement in the placement process and consideration of more than one placement site
- a clearly articulated learning contract that identifies mutual rights, responsibilities, and expectations
- a service contract with the agency that defines the limitations of the intern's role
- liability insurance
- the prior knowledge clause
- assurance of work and field site safety
- assumption of risk as limited to ordinary risk
- employer-employee-independent contractor relationship
- compensation: stipend, scholarship, taxable/taxfree
- negligence
- malpractice
- implication of federal funds and related statutory and regulatory requirements
- use of college work-study funds for interns
- grievance processes
- informed consent in accepting an internship
- respecting the prerogative and obligations of the institution
- responsibility to confront unethical/illegal behaviors
- public representation of self and work

where you have been offered a paid field experience by the agency's director. The site is out of state and is not one of your campus' regularly used sites. Your family is relocating to that state because of financial reasons, and having a paid internship in that area of the state would be of great help financially and prevent greater hardship on your family. You doubt that the field placement coordinator for your academic program is aware of the improprieties at the agency. And you are questioning whether to inform the campus about what you know or be silent and help your family.

You might be wondering how you have managed to survive for so long without knowing about these issues! This is exactly how our students feel after studying them. Chances are that your basic values have held you in good stead. However, the list can be overwhelming at first, and even after several readings. Most likely, there are many issues about which you never heard and many that seem remotely familiar. Reading through them is a good start. Now that you are becoming familiar with the language of ethics, even if you do not yet know everything it entails, it is a good time to go ahead and identify the dilemmas that already have surfaced in your internship. Are you satisfied with how you responded to them?

Deciding Dilemmas

An ethical dilemma is a difficulty that you may in fact face in the course of your internship. You are no stranger to facing difficulties; you have managed to confront crises in your growth and in situations that may have been painful and exasperating. Like the challenges you faced in the past, an ethical difficulty can become a problem if not managed effectively. One way of dealing with an ethical issue is to anticipate it. A useful and effective way of anticipating it is to rehearse in your mind the best possible response (how you would like to respond to it), the worst possible response (your worst nightmare), and a more realistically based response (found somewhere between the two extremes). However, there are many situations for which this framework will not work, and for these, you need a decision-making guide.

There are many guides from which to choose, and nearly every introductory text in the field of human service education includes a chapter on professional issues and a decision-making model (see, e.g., M. S. Corey & Corey, 1998; Neukrug, 1994; Schram & Mandell, 1997; Woodside & McClam, 1997). The model we offer is one that we use when teaching professional issues. It is based in part on the Steps to Creating Change, incorporates adaptations of other models (Close & Meier, 1995; G. Corey et al., 1998), and includes a step to use specifically when the issue you uncover is an ethical or legal one.

NINE STEPS TO DECIDING DILEMMAS

1. Name the Problem Collect as much information as you can about the situation. Clarify the conflict. Is it moral? Professional? Ethical? Legal? Given that there are no right or wrong answers to the situation, anticipate ambiguity and challenge yourself to consider the problem from multiple perspectives.

2. Narrow the Focus Once you have gathered as much information as is reasonable, list the issues you are confronting. Some are more important than others. Describe the critical issues and players; discard the unimportant ones.

3. Consult the Codes Review the guidelines/codes/standards of your profession and the policies and regulations of your agency, as well as the related laws, to determine whether possible solutions are suggested. Identify aspects of the codes that apply. How compatible are your personal values with those of the profession? What rationales support those areas of conflict between your personal values and ethics and those of the profession?

4. Consult with Colleagues Consult with informed colleagues for other ways to consider the problem. This process can be especially helpful in thinking through the circumstances and information and in identifying possible gaps and issues not already considered. Given the responsibility to make a reasoned decision, consulting with colleagues is one way to "act in good faith" and test your justifications. Choose your colleagues wisely.

5. Determine the Goals One of the most important things to think through is what you hope to see as a result of action being taken about the attitudes, behaviors, or circumstances in question. Question your motives carefully and repeatedly. Is your client's voice heard in the goals you want? Talk with a colleague about your goals for a resolution and whether or not there may be motives on your part of which you are not aware. Choose your colleague wisely.

6. Brainstorm the Strategies Identify all possible courses of action, including the absurd. Some may prove useful, though unorthodox. Consider the client's perspective as well. Is the client's voice being heard in your list of options? Discuss options with others. Choose your colleagues wisely.

7. Consider the Consequences Think about the consequences of each strategy for all involved in the situation. Importantly, whatever the plans, they must be thoughtfully assessed. Your task is to identify consequences from various perspectives and to question each of the consequences you identify. Remember to include the clients' perspective among those you consider.

8. Consult the Checklist Use the following checklist as a framework to evaluate potential areas of ethical and legal misconduct. The questions are based in part on a model of ethical decision making that identifies six fundamental principles of moral behavior: autonomy (self-determination), beneficence (in the best interest of the client), nonmaleficence (to do no harm), justice (fairness to all), fidelity (honest promises and honored commitments), and veracity (being truthful). This model includes such qualities of ethical acts as universality, morality, and reasoned and principled behaviors (G. Corey et al., 1998; Kitchener, 1984; Pollock, 1998).

- *Is the action in the best interest of the client?* Consider the six fundamental principles of moral behavior.
- *Does the action violate the rights of another person?* Consider constitutional rights as well as your duty to justice.
- *Does the action involve treating another person only as a means to achieve a self-serving end?* Consider the end-in-itself motive and the utilitarian perspective.
- *Is the action under consideration legal? Is it ethical?* Consider your legal, civic, ethical duties, and the components of an ethical act.
- *Does the action create more harm than good for those involved?* Consider the principles of nonmaleficence and beneficence.
- *Does the action violate existing policies, regulations, procedures, or professional standards?* Consider the duty to one's professional role.
- *Does the action promote values in culturally affirming ways?* Consider the principles of nonmaleficence and beneficence and the duty to care.

9. Decide With Care Consider carefully the information you have. The more obvious the dilemma, the clearer the course of action; the more nebulous the dilemma, the more difficult the choice. Though hindsight may teach you differently, the best decision in these circumstances is a well-reasoned decision with which you can live.

SUMMARY

The rewards of this long journey in learning are finally realized when you enter the competence stage. Although hardly without its challenges, this stage is one in which you can indulge yourself and enjoy the feelings of finally reaching your goals. The transformations in development that you have experienced and the sense of empowerment that you have developed are evident not only to you in how you feel and go about your work but to those around you as well. As you continue to develop your competencies and your sense of professional identity, you are fast approaching the last mile of your journey. It is about to begin.

For Further Reflection

1. *Thinking about fulfillment.* Take a moment to think about just how fulfilled you are with your internship. What aspects of your experience give you a sense of fulfillment? Which tend to interfere with it? You might want to try this simple, yet informative, exercise. On a piece of paper, jot down all the factors you can think of (including the ones mentioned in this chapter) that are important to your internship (positive and negative). Then, next to each one, indicate the amounts of "too much," "too little," or "just right." Chances are that what's just right leaves you with the greatest feelings of satisfaction.
2. *Thinking about professionalism.* Read over the following questions and think about the ways in which your perspectives about professionalism have changed during the internship.
   - What is your operating definition of professional? What behaviors and values do you associate with being professional?
   - What professional behaviors are the norm for your field site? What behaviors contradict the norms? What values are reflected in both instances?
   - Think about someone you met in the course of your internship who meets your definition of professional. What is it about this individual that you find particularly admirable?
   - How do you see yourself measuring up to your definition of professional? (Be as specific as you can be.)
3. *Thinking about your personal system of ethics.* This is an exercise in reflection for journal writing. In thinking about the personal system of ethics that you live by, keep in mind that yours might be similar to another person's in some ways, but you will always find areas of differences. Your system of ethics is unique, regardless of the shared ethics that your profession mandates.

- How do you go about confronting ethical decisions? What words best describe your style: Thoughtful? Impulsive? Convenient? Pressured? Expedient? Others?
- What values define your personal set of ethics? Your professional set of ethics? Are there areas of potential conflict between the two?
- How consistent is your behavior with these values—personally and professionally?
- Have either your values or ethics changed since you began working in the field of human services? If so, in what ways? What has influenced your ways of thinking? Behaving?
- What areas of your ethical conduct have caused you concern in the past? To what areas do you want to pay particular attention in the future?
- When you think of professional loyalties, how do you prioritize them? Where do you experience conflicts between them? How do you deal with these conflicts?
- What is your style of accepting responsibility? When you are responsible for a situation that results in consequences, how do you typically respond? In what ways does your style of reacting help you better yourself? In what ways does it hold you back?
- When you are confronted with an ethical decision, how important are the following considerations?

  In what ways does your decision affect others?

  Does your decision hurt others?

  Would you make the same decision if you were on the other side of the issue?

  How proud are you of your decision?

  How proud would your family be?
- What are the sources of your systems of values and ethics? How did you come by them? If you could change those circumstances, would you? In what ways? Why?
- What ethical dilemma do you most remember from childhood? Why do you think that is? How did you deal with it then? Would you deal with it differently now? What life lesson did it give you?

For Further Exploration

READINGS

Baird, B. N. (1996). *The internship, practicum, and field placement handbook.* Upper Saddle River, NJ: Prentice Hall.

Comprehensive and especially useful to graduate students in the helping professions. Covers a wide range of topics relevant to aspects of an internship. Strong focus on the clinical/counseling dimension of human service work.

Brill, N. I. (1998). *Working with people* (5th ed.). New York: Longman.

Thoughtful consideration of a variety of professional, moral, and ethical issues.

Corey, G., Corey, M., & Callanan, P. (1998). *Issues and ethics in the helping professions* (5th ed.). Pacific Grove, CA: Brooks/Cole.

A seminal text that belongs in the office of every graduating student entering the field of human services. An essential text for all courses that focus on the legal, ethical, or professional dimensions of the helping professions.

Goldstein, M. B. (1990). Legal issues in combining service and learning. In J. C. Kendall & Associates (Eds.), *Combining service and learning* (Vol. 2, pp. 39–60). Raleigh, NC: National Society for Experiential Education.

Offers a basic guide to the legal issues involved in combining education and learning in the community.

Jackson, R. (1997). Alive in the world: The transformative power of experience. *N.S.E.E. Quarterly*, 22(3), pp. 1, 24–26.

Explores what it is about experiential education that "gives experience the power to transform."

PROFESSIONAL ORGANIZATIONS

The following organizations can provide information about the professional standards specific to each of their disciplines in human services.

1. American Academy of Criminal Justice Sciences
   Northern Kentucky University
   402 Nunn Hall
   Highland Heights, KY 41099-5998
   Telephone: (606) 572-5634
2. American Counseling Association
   5999 Stevenson Avenue
   Alexandria, VA 22304
   Telephone: (703) 823-9800
3. American Psychological Association
   1200 17th Street, N.W.
   Washington, DC 20036
   Telephone: (202) 955-7600
4. American Association for Marriage and Family Therapy
   1133 15th Street
   Washington, DC 20005-2710
   Telephone: (202) 452-0109
5. National Association of Social Workers
   750 First Street, N.E.
   Suite 700
   Washington, DC 20002-4241
   Telephone: (800) 638-8799
6. National Organization for Human Service Education
   Home page: http://www.nohse.com

References

American Psychological Association. (1995). *Ethical principles of psychologists and code of conduct.* Washington, DC: Author.

Baird, B. N. (1996). *The internship, practicum, and field placement handbook: A guide for the helping professions.* Upper Saddle River, NJ: Prentice Hall.

Chiaferi, R., & Griffin, M. (1997). *Developing fieldwork skills.* Pacific Grove, CA: Brooks/Cole.

Chickering, A. W. (1969). *Education and identity.* San Francisco: Jossey-Bass.

Close, D., & Meier, N. (1995). *Morality in criminal justice.* Belmont, CA: Wadsworth.

Collins, D., Thomlison, B., & Grinnell, R. M. (1992). *The social work practicum: A student guide.* Itasca, IL: F. E. Peacock.

Collins, P. (1993). The interpersonal vicissitudes of mentorship: An exploratory study of the field supervisor-student relationship. *Clinical Supervisor, 11*(1), 121–136.

Corey, G., Corey, M., & Callanan, P. (1998). *Issues and ethics in the helping professions* (5th ed.). Pacific Grove, CA: Brooks/Cole.

Corey, M. S., & Corey, G. (1998). *Becoming a helper* (3rd ed.). Pacific Grove, CA: Brooks/Cole.

Dreyfus, S. E., & Dreyfus, H. L. (1980). *A five stage model of the mental activities involved in directed skill acquisition* (Unpublished Report F49620-79-C-0063). Air Force Office of Scientific Research (AFSC), University of California, Berkeley.

Erikson, E. H. (1963). *Childhood and society.* New York: Norton.

Goldstein, M. B. (Ed.). (1990). Legal issues in combining service and learning. In J. C. Kendall & Associates (Eds.), *Combining service and learning* (Vol. 2, pp. 39–60). Raleigh, NC: National Society for Experiential Learning.

Gordon, G. R., & McBride, R. (1996). *Criminal justice internships: Theory into practice* (3rd ed.). Cincinnati, OH: Anderson Publishing.

Hersey, P., & Blanchard, K. (1982). *Management of organizational behavior: Utilizing human resources* (4th ed.). Upper Saddle River, NJ: Prentice Hall.

Huber, R. M. (1971). *The American idea of success.* New York: McGraw-Hill.

Kanter, R. M. (1977). *Men and women of the corporation.* New York: Basic Books.

King, M. A. (1988). *Toward an understanding of the phenomenology of fulfillment in success.* Unpublished doctoral dissertation. University of Massachusetts, Amherst.

Kitchener, K. S. (1984). Intuition, critical evaluation and ethical principles: The foundation for ethical decisions in counseling psychology. *The Counseling Psychologist, 12*(3), 43–45.

Kolb, D. A. (1984). *Experiential learning: Experience as the source of learning and development.* Upper Saddle River, NJ: Prentice Hall.

Lamb, D. H., Baker, J. M., Jennings, M. L., & Yarris, E. (1982). Passages of an internship in professional psychology. *Professional Psychology, 13*(5), 661–669.

Martin, M. L. (Ed.). (1991). *Employment setting as practicum site: A field instruction dilemma.* Dubuque, IA: Kendall/Hunt.

Mozes-Zirkes, S. (1993, July). Mentoring integral to science, practice. *APA Monitor,* p. 34.

National Organization for Human Service Education. (1996). Special feature: Ethical standards of human service professionals. *Human Service Education, 16*(1), 11–17.

Neukrug, E. (1994). *Theory, practice and trends in human services: An overview of an emerging profession.* Pacific Grove, CA: Brooks/Cole.

Pollock, J. M. (1998). *Ethics in crime and justice: Dilemmas and decisions* (3rd ed.). Belmont, CA: Wadsworth.

Royse, D., Dhooper, S. S., & Rompf, E. L. (1996). *Field instruction: A guide for social work students* (2nd ed.). New York: Longman.

Schram, B., & Mandell, B. R. (1997). *An introduction to human services* (3rd ed.). New York: Macmillan.

Schultz, M. (1992). Internships in sociology: Liability issues and risk management measures. *Teaching Sociology, 20,* 183–191.

Tentoni, S. C. (1995). The mentoring of counseling students: A concept in search of a paradigm. *Counselor Education and Supervision, 35*(1), 32–41.

Tryon, G. S. (1996). Supervisee development during the practicum year. *Counselor Education and Supervision, 35*(4), 287–294.

Wilson, S. J. (1981). *Field instruction: Techniques for supervisors.* New York: Free Press.

Woodside, M., & McClam, T. (1997). *An introduction to human services* (3rd ed.). Pacific Grove, CA: Brooks/Cole.

APPENDIX

ETHICAL STANDARDS OF HUMAN SERVICE PROFESSIONALS

NATIONAL ORGANIZATION FOR HUMAN SERVICE EDUCATION
COUNCIL FOR STANDARDS IN HUMAN SERVICE EDUCATION

Preamble

Human services is a profession developing in response to and in anticipation of the direction of human needs and human problems in the late twentieth century. Characterized particularly by an appreciation of human beings in all of their diversity, human services offers assistance to its clients within the context of their community and environment. Human service professionals, regardless of whether they are students, faculty or practitioners, promote and encourage the unique values and characteristics of human services. In so doing human service professionals uphold the integrity and ethics of the profession, partake in constructive criticism of the profession, promote client and community well-being, and enhance their own professional growth.

The ethical guidelines presented are a set of standards of conduct which the human service professional considers in ethical and professional decision making. It is hoped that these guidelines will be of assistance when the human service professional is challenged by difficult ethical dilemmas. Although ethical codes are not legal documents, they may be used to assist in the adjudication of issues related to ethical human service behavior.

Human service professionals function in many ways and carry out many roles. They enter into professional-client relationships with individuals, families, groups and communities who are all referred to as "clients" in these standards. Among their roles are caregiver, case manager, broker, teacher/educator, behavior changer, consultant, outreach professional, mobilizer, advocate, community planner, community change organizer, evaluator and administrator. The following standards are written with these multifaceted roles in mind.

The Human Service Professional's Responsibility to Clients

STATEMENT 1 Human service professionals negotiate with clients the purpose, goals, and nature of the helping relationship prior to its onset as well as inform clients of the limitations of the proposed relationship.

STATEMENT 2 Human service professionals respect the integrity and welfare of the client at all times. Each client is treated with respect, acceptance and dignity.

STATEMENT 3 Human service professionals protect the client's right to privacy and confidentiality except when such confidentiality would cause harm to the client or others, when agency guidelines state otherwise, or under other stated conditions (e.g., local, state, or federal laws). Professionals inform clients of the limits of confidentiality prior to the onset of the helping relationship.

STATEMENT 4 If it is suspected that danger or harm may occur to the client or to others as a result of a client's behavior, the human service professional acts in an appropriate and professional manner to protect the safety of those individuals. This may involve seeking consultation, supervision, and/or breaking the confidentiality of the relationship.

STATEMENT 5 Human service professionals protect the integrity, safety, and security of client records. All written client information that is shared with other professionals, except in the course of professional supervision, must have the client's prior written consent.

STATEMENT 6 Human service professionals are aware that in their relationships with clients, power and status are unequal. Therefore they recognize that dual or multiple relationships may increase the risk of harm to, or exploitation of, clients, and may impair their professional judgment. However, in some communities and situations it may not be feasible to avoid social or other nonprofessional contact with clients. Human service professionals support the trust implicit in the helping relationship by avoiding dual relationships that may impair professional judgment, increase the risk of harm to clients or lead to exploitation.

STATEMENT 7 Sexual relationships with current clients are not considered to be in the best interest of the client and are prohibited. Sexual relationships with previous clients are considered dual relationships and are addressed in Statement 6 (above).

STATEMENT 8 The client's right to self-determination is protected by human service professionals. They recognize the client's right to receive or refuse services.

STATEMENT 9 Human service professionals recognize and build on client strengths.

The Human Service Professional's Responsibility to the Community and Society

STATEMENT 10 Human service professionals are aware of local, state, and federal laws. They advocate for change in regulations and statutes when such legislation conflicts with ethical guidelines and/or client rights. Where laws are harmful to individuals, groups or communities, human service professionals consider the conflict between the values of obeying the law and the values of serving people and may decide to initiate social action.

STATEMENT 11 Human service professionals keep informed about current social issues as they affect the client and the community. They share that information with clients, groups and community as part of their work.

STATEMENT 12 Human service professionals understand the complex interaction between individuals, their families, the communities in which they live, and society.

STATEMENT 13 Human service professionals act as advocates in addressing unmet client and community needs. Human service professionals provide a mechanism for identifying unmet client needs, calling attention to these needs, and assisting in planning and mobilizing to advocate for those needs at the local community level.

STATEMENT 14 Human service professionals represent their qualifications to the public accurately.

STATEMENT 15 Human service professionals describe the effectiveness of programs, treatments, and/or techniques accurately.

STATEMENT 16 Human service professionals advocate for the rights of all members of society, particularly those who are members of minorities and groups at which discriminatory practices have historically been directed.

STATEMENT 17 Human service professionals provide services without discrimination or preference based on age, ethnicity, culture, race, disability, gender, religion, sexual orientation or socioeconomic status.

STATEMENT 18 Human service professionals are knowledgeable about the cultures and communities within which they practice. They are aware of multiculturalism in society and its impact on the community as well as individuals within the community. They respect individuals and groups, their cultures and beliefs.

STATEMENT 19 Human service professionals are aware of their own cultural backgrounds, beliefs, and values, recognizing the potential for impact on their relationships with others.

STATEMENT 20 Human service professionals are aware of sociopolitical issues that differentially affect clients from diverse backgrounds.

STATEMENT 21 Human service professionals seek the training, experience, education and supervision necessary to ensure their effectiveness in working with culturally diverse client populations.

The Human Service Professional's Responsibility to Colleagues

STATEMENT 22 Human service professionals avoid duplicating another professional's helping relationship with a client They consult with other professionals who are assisting the client in a different type of relationship when it is in the best interest of the client to do so.

STATEMENT 23 When a human service professional has a conflict with a colleague, he or she first seeks out the colleague in an attempt to manage the problem. If necessary, the professional then seeks the assistance of supervisors, consultants or other professionals in efforts to manage the problem.

STATEMENT 24 Human service professionals respond appropriately to unethical behavior of colleagues. Usually this means initially talking directly with the colleague and, if no resolution is forthcoming, reporting the colleague's behavior to supervisory or administrative staff and/or to the professional organization(s) to which the colleague belongs.

STATEMENT 25 All consultations between human service professionals are kept confidential unless to do so would result in harm to clients or communities.

The Human Service Professional's Responsibility to the Profession

STATEMENT 26 Human service professionals know the limit and scope of their professional knowledge and offer services only within their knowledge and skill base.

STATEMENT 27 Human service professionals seek appropriate consultation and supervision to assist in decision-making when there are legal, ethical or other dilemmas.

STATEMENT 28 Human service professionals act with integrity, honesty, genuineness, and objectivity.

STATEMENT 29 Human service professionals promote cooperation among related disciplines (e.g., psychology, counseling, social work, nursing, family and consumer sciences, medicine, education) to foster professional growth and interests within the various fields.

STATEMENT 30 Human service professionals promote the continuing development of their profession. They encourage membership in professional associations, support research endeavors, foster educational advancement, advocate for appropriate legislative actions, and participate in other related professional activities.

STATEMENT 31 Human service professionals continually seek out new and effective approaches to enhance their professional abilities.

The Human Service Professional's Responsibility to Employers

STATEMENT 32 Human service professionals adhere to commitments made to their employers.

STATEMENT 33 Human service professionals participate in efforts to establish and maintain employment conditions which are conducive to high quality client services. They assist in evaluating the effectiveness of the agency through reliable and valid assessment measures.

STATEMENT 34 When a conflict arises between fulfilling the responsibility to the employer and the responsibility to the client, human service professionals advise both of the conflict and work conjointly with all involved to manage the conflict.

The Human Service Professional's Responsibility to Self

STATEMENT 35 Human service professionals strive to personify those characteristics typically associated with the profession (eg., accountability, respect for others, genuineness, empathy, pragmatism).

STATEMENT 36 Human service professionals foster self-awareness and personal growth in themselves. They recognize that when professionals are aware of their own values, attitudes, cultural background, and personal needs, the process of helping others is less likely to be negatively impacted by those factors.

STATEMENT 37 Human service professionals recognize a commitment to lifelong learning and continually upgrade knowledge and skills to serve the populations better.

CHAPTER 11

The Last Mile: The Culmination Stage

I've been taking in as much information as I can before "it's all over" and have been concerned with career goals. The past couple of weeks have just been filled with an overwhelming amount of anxieties and mixed emotions.

STUDENT JOURNAL ENTRY

As incredible as it must seem to you at times, the end is in sight. For some of you, the beginning of your internship seems like yesterday; for others, it seems like years ago; still others report both feelings. It is a time to look forward to the future and to be proud of what you have accomplished. So why don't you always feel clear and positive about the experience and about ending it?

In our experience, interns approaching the end of their field experience report many different feelings. There is indeed pride and even feelings of mastery. Others report relief and anticipation of freedom. However, we also hear about sadness, anger, loss, and confusion. In a survey of psychotherapy interns, Robert Gould (1978) found that many of them reported increased anxiety and depression as the end approached as well as decreased effectiveness. The interns in his study also reported feeling moodier. That has been our experience as well; you would not be unusual if you found yourself experiencing several—or even all—of these feelings, simultaneously or sequentially, with your emotional landscape shifting by the hour. What is going on here?

For one thing, you have a lot on your plate right now. Your internship is ending, and you may or may not be ready. The calendar says the internship is over, but you may feel as though you have just gotten started, especially if it took some time to hit your stride (Suelzle & Borzak, 1981). On the other hand, you may feel as though you have been finished for a while. In addition, you may or may not have completed the work you set out for yourself. Projects may not be finished; clients may not be quite where you would like them to be.

The web of relationships that has been the social context of your internship is changing yet again. Many relationships are ending; others will be redefined as you leave the role of intern. Endings are part of most relationships and a necessary part of helping relationships (Brill, 1998). The more the relationships mean to you, though, the harder it will be to end them. Some interns find that the impact of relationships actually reaches its peak as the relationships near their end (Gould, 1978):

> Now I am really sad. I loved my internship and don't want to lose the relationships I've formed here. I'm trying to make plans with everyone so we can keep in touch.

The external context of your life is shifting as well. There may be new demands on your time and energy as you feel the pressure of endings and beginnings (Suelzle & Borzak, 1981). Your attention may be pulled toward papers and final projects for your internship seminar as well as to papers and exams in other classes. Your summer or holiday plans may need to be finalized. If you are graduating and have not yet found a job, your job search is occupying more and more time and attention.

Of course, your internship is not the only thing that is ending. Your seminar class, the semester, the school year, even your college career may all be drawing to a close. That's a lot of endings, a lot of beginnings, and a lot of good-byes. Good-byes are never easy, and for some they can be very stressful. The prospect of beginning in a new school, a new town, a new job, or all of the above is daunting as well as exciting.

That is why we call this the Culmination stage; everything is winding down. Another stage, another crisis, and another set of risks and opportunities await you. Like other critical junctures discussed in this book, this one is normal and so are the concerns and issues associated with it. Endings are a necessary part of anyone's development (Kegan, 1982). Normal does not mean easy, though. Remember the distinction between difficulties and problems discussed in Chapter 9? You could think of endings as difficulties that can either be solved, resolved, or aggravated into problems (Watzlawick, Weaklund, & Fisch, 1974).

There are opportunities here for you to grow, to develop new insights, and to learn more about separation and moving on, but there are also risks. It is natural to want to avoid the conflicting feelings that can surface at this time, and that avoidance can take many forms, including lateness or absence, devaluing the experience (suddenly it doesn't seem all that great anymore), or putting on rose-colored glasses and forgetting all the struggles you had and may still be having. The risk in these behaviors is that you can turn a difficulty into a problem. Lateness and absence will not help anything and will create new problems for you. Becoming out of touch with your own emotions may make you less available to clients. You also risk losing the opportunity to learn how to face the issues and feelings associated with endings. Finally, interns who do not face these issues often report a hollow feeling as they leave their placements, even though it may have been a good experience in many ways.

SEIZING THE OPPORTUNITIES

The rest of this chapter is designed to help you minimize the risks, meet the challenges, and maximize the opportunities. Making time to process your experience, taking a new self-inventory, and facing the tasks of ending are all essential components in this effort.

As your internship ends, you may feel more than ever the pull to do more and more. There may be a lot that can or needs to be finished, but becoming too involved in a frenzy of activity can also be a way to avoid facing your feelings. Once again, we remind you—and it is especially important now—that you need time and energy for reflection. You will have to protect that time more jealously than ever, as the external pressures mount and the temptation to avoid your feelings grows stronger.

Taking a Self-Inventory

As you head into this phase of your internship, take some time to think about yourself. All these endings and beginnings are likely to tap a variety of emotional issues. Some of the issues you thought about in Chapters 2 and 3 may resurface or surface for the first time. You may also discover some new issues. Separation and loss can leave people with some unfinished emotional business. Think back on the experiences of separation and loss in your life. Going away to camp or to college, having close siblings leave home, moving, divorce, ending an intimate relationship, and being fired from a job are examples of separation experiences you may have had. Those experiences are never easy, and for some people in some circumstances, they can be traumatic. Perhaps some of the hurts from those experiences have not healed. Perhaps you wish you had behaved differently or that others had.

Now think about how you say good-bye and how people have said good-bye to you. Some people just leave and don't say a word; others write long letters or schedule good-bye lunches or dinners. The way you say good-bye probably depends somewhat on the nature of the relationship that is ending, but perhaps you can discern some patterns in how you approach this task. Remember the discussion of dysfunctional patterns in Chapter 2? This is an area in which many people have those patterns, and they can surface in an internship. Here is one we have seen many times:

> Saying good-bye to someone I don't care that much about is pretty easy. When it's someone I'm really invested in, though, I don't handle it well. Usually what I do is get really busy. I keep promising myself to go to lunch, or dinner, or something with the person, but I never seem to make the time. Then all of a sudden there is no time, and I end up saying a hurried good-bye. I know I have hurt people's feelings that way.

Scott Haas (1990), in a book on his psychology internship, notices himself falling into a similar pattern as the end of his internship approaches:

> I create obstacles to avoid thinking about the end. I distract myself: I make lists. I ruminate about minor inconveniences (like wondering for days whether the gas company will correct their bill). I develop new projects and interests. I go shopping, and then in the store can't remember why I ever wanted the thing I'm about to buy. When all else fails, I pretend there is no end. I'm just imagining it: it really isn't happening. (p. 171)

Another area that may be touched by the culmination stage is your feelings about the issues of competence. You may recall from Chapter 2 that this is one of the components of your psychosocial identity. Some interns report that they feel competent for the very first time during their internship, and that is very hard to leave behind. If your sense of competence is tied to success, then you may be trying to get in one more success before you leave, pursuing perfection in a project or one more milestone with a client. Richard Schafer (1973) believes that many helping professionals strive for a "sense of goodness." Although achievement is part of this sense, another component is feeling like a good person, regardless of one's faults. Gould (1978) points out that this particular sense of goodness can be difficult for an intern to achieve. Many people entering the helping professions, he says, want and need to be liked and seek approval from their clients. Their sense of competence, and goodness, is tied to client success or to some other form of success at their site.

Being aware of how you deal with endings and knowing where you are vulnerable will help you be open and sensitive to the experience of your clients. You can avoid blaming others for your feelings and experiences. Finally, you can make conscious efforts to avoid the pitfalls.

The Tasks of Ending

According to Naomi Brill (1998), there are three tasks associated with what we are calling the Culmination stage. They are not necessarily to be completed in sequential order; in fact, you will surely see that they are interconnected and are usually dealt with simultaneously. One important task is that unfinished business must be identified and dealt with. These are issues with clients, supervisors, co-workers, and yourself that have been present for some time but that often take on added urgency as the internship draws to a close (Shulman, 1983). Next, you must identify your feelings and find a safe place to express them. You may find strong feelings about supervisors, clients, peers, and others coming to the surface. It may or may not be appropriate to express these feelings directly to the person who has engendered them. At the very least, you will want to find a place where you can express them to someone else and say whatever you want to say before worrying about just what to say to the people involved. Finally, you need to plan for the future.

These three tasks can be applied to the work you are doing, the people you are interacting with, and the placement site as a whole. In some cases, the tasks can be attended to informally, and each intern will do them differently. In other cases, however, we are going to suggest that you be formal and structured about it.

In writing about ending an internship, some authors have discussed the need for rituals (Baird, 1996; Collins, Thomlison, & Grinnell, 1992). That may sound like a strange term to you, even calling up religious or pagan scenes. However, any formal way of marking an event or passage (such as a going-away luncheon) can be considered a ritual. In this case, rituals can help provide a sense of completion or closure. Benjamin Baird (1996), whose writings about the end of an internship are especially thoughtful, discusses several benefits of rituals in the closing phase. Rituals can add a sense of specialness to the ending; they can help you recall what was significant and important. They can also help ease the transition by connecting the past to the future. Finally, rituals can create a formal, structured opportunity to experience and express emotions that might otherwise be repressed.

FINISHING THE WORK

As you enter the last weeks of your internship, you need to think about what tasks remain and what you want to and can do about them. The nature of the tasks makes a difference here. If you have been given a series of small concrete projects, such as a report to write or an event to plan, and they are complete or near completion, then ending the internship will be somewhat easier (Suelzle & Borzak, 1981). If you are part of a large project that will not end until after you have gone, then feeling some closure about your involvement can be harder. Perhaps there is a component of the project you can complete or a summary of your work that you can write. Finally, you may be involved in a complex project that must be completed before you leave. Perhaps, for example, you have done some research for your supervisor and have agreed to summarize your findings. There may be the temptation to read one more article or interview one more person; in some cases, there is always more research you can do.

Planning for the future also means considering the nature of your involvement, if any, with the agency and its work after your internship is over. Some interns are offered part-time work, relief work, or even full-time jobs. That is great when it happens; just make sure that accepting is really your best option and not just a way to avoid bringing things to an end. In other cases, you may want to come back for a visit or to see and help with an event you have been working on. Coming back for a visit can be fine depending on the nature of your clients (this issue will be discussed more in another section). Seeing a project or event you have helped plan come to fruition can be wonderful as well. Just take care not to promise more than you can deliver. If the agency is counting on you and you get caught up in your life and don't follow through, you leave a bad feeling about you and possibly about your school.

SAYING GOOD-BYE TO YOUR CLIENTS

Termination is the word used to describe the ending of a therapeutic relationship. To some of you, this is a familiar term. You may have studied it in class, and if your agency does one-to-one counseling, you may have heard the term there as well. For others the term may be new, somewhat strange, and harsh sounding, especially given its recent use in the movies and on television. Not all of you are doing counseling or psychotherapy in your internship, but if you are doing any sort of direct work with clients, you need to think carefully about ending your work with them and saying good-bye, regardless of the label you apply to the process.

Furthermore, as an intern you are saying good-bye under unusual circumstances. Usually when a client-worker relationship ends, it is because the work is done or the client decides to end it. However, you are dealing with what are called *forced terminations* (Gould, 1978). It is the calendar that dictates the ending. The work is not necessarily over, and the clients are not necessarily ready to stop or start over with someone else. Ending relationships with clients is bound to engender emotional reactions for them and for you. Dealing with those feelings and ending well is an important part of the internship. For some of you, it will be relatively easy. For others, it may be one of the biggest challenges you face.

Part of the reason our interns have such varied experiences in coming to closure with clients is that there are such great variations in the internship sites. Some sites do a lot of one-to-one counseling. Others, such as adolescent shelters, do some, but in the context of daily living and recreation. Other sites do none at all. There is also great variation in the characteristics of the clients. Some client populations are much more emotionally vulnerable and may take your departure that much harder. Some clients are more autonomous than others. Clients who drop into a senior services center, for example, or a town recreation department are probably functioning pretty well on their own. They appreciate you, but they really don't need you. Finally, some clients are simply more aware of your leaving than others. One intern, who was working with a number of clients who had Alzheimer's disease, reported that she had to tell each client over and over that she would be leaving. Many of them did not remember her from day to day, although they seemed glad to see her. Leaving was difficult for her, but probably not for them.

In any case, there are four important issues to think about regarding closure with your clients: (a) you need to decide when and how to tell them, (b) you need to identify and take care of any unfinished business with them, (c) you need to deal with your feelings and theirs about your leaving, and (d) you need to plan for their future needs.

Timing and Style

There are many different theories and opinions about when and how to tell clients you are leaving. Because internship sites and client populations vary so greatly, you must look to your site, your supervisor, and your instructor to guide you in this area; no prescriptions we could make would apply to every intern, or even to most of them. Some agencies recommend that you tell clients right away that you will be leaving at the end of the term or the year; in fact, some agencies tell the clients for you. Some clients are used to seeing interns come and go; as soon as you say you are an intern, they know you will not be there for very long. Other agencies recommend that you begin discussion of termination 1 week, 2 weeks, or more in advance. Some recommend that you not mention it until right before you go.

There are also a variety of methods of discussing termination, from individual conferences, to group meetings, to letters, or any combination of these approaches. If you do not know how your agency handles termination, you need to take the initiative to ask. Depending on your needs and attitudes about ending, you may feel that the agency's policy is too casual or forces you to focus on something that does not seem like a big deal. Remember that the clients are the primary focus; the decision needs to be made based on what will work best for them.

Unfinished Business

As you approach the end of your relationship with clients, it is a good time to reflect, certainly in private and perhaps with your clients, on their progress. If they had specific goals, this is the time to review them. Even if they did not, your memory and your journal are good sources for remembering what your clients were like when you first arrived, as well as what your relationship with them was like. Both clients and workers can become so caught up in the problems and challenges of the present that they don't think much about the change that has already occurred. Add to that the difficult feelings that termination can bring and you can see why it is important to take time to reflect on the positive.

It is equally important, though, to be clear with yourself and with clients about the work that remains. All of this is especially important if the client is going to continue at the agency, which is usually the case. You may want to use some of your supervision time to review each of your clients with your supervisor, or you may want to discuss them as a group. You will want to talk about how you feel about each of these terminations, as well as how you think each client is going to react. Baird (1996) discusses a Termination Scale developed by Fair and Bressler (1992) that has 55 items covering emotional responses and planning. You may want to make use of this resource.

Dealing With Feelings

In many cases, the termination process can be an emotional one for everyone involved. Clients may be especially vulnerable. For them, your leaving may recall echoes of their past experiences of separation and loss. For many of these clients—especially those who have been subjected to abuse, neglect, or parents with substance abuse problems—separations have been arbitrary and unpredictable, such as when a parent abandons a child or an alcoholic parent goes on a drinking binge and returns days later. For some people, separation is associated with anger and hysteria, as in the loud and even violent arguments that can lead to the disruption or ending of a marriage or other intimate relationship. You may also find that clients have had particularly traumatic separations, as in the death of a parent, sibling, or close friend. Finally, many of your clients will have worked with several agencies, workers, or several residential settings and have had many termination experiences. If some or all of these experiences have been painful, this termination may bring up those feelings again (Baird, 1996), and you are right to be concerned.

> *"Why do you have to go?" is probably the most difficult [question]. Many of these children have had people who come and go in their lives and I don't want them to look at me as one of those people.*
>
> STUDENT JOURNAL ENTRY

Of course, termination is also a learning opportunity. Clients can learn that good-byes, while often painful, do not have to be traumatic, angry, or hysterical.

Many interns report feeling nervous and apprehensive about discussing termination with clients because they are afraid of what the reaction might be. Well, as you can imagine, the variety of personalities in your clients and the separation experiences they have had will result in an equal variety of reactions to the news that you are leaving. Some may feel abandoned. After all, you are leaving because of your school calendar, not because they don't need you, so they may wonder how much you really cared in the first place (Baird, 1996; Stanziani, 1993). Some may feel angry and believe their trust has been betrayed, even if you have been clear from the beginning that you will be leaving at the end of the semester. It is important to remember that whatever reaction you receive may be only part of the picture (Penn, 1990). Termination, like the end of any significant relationship, usually creates feelings of anger, sadness, and appreciation. Your client may be expressing only one of those feelings, and indeed that may be the only feeling he or she can access, but it is important for you to remain open to the other feelings as well, and perhaps to help your client be open to them, too.

Sometimes these feelings are expressed indirectly, not unlike your own feelings about ending the internship (Penn, 1990). Some clients may be genuinely indifferent to your leaving. Others will feign indifference or be indifferent because their real feelings are too hard to accept. Another indirect way of dealing with termination issues is to demand that termination come immediately. In these cases, once you tell clients you will be leaving, they may stop coming to the agency or in residential settings ask for another worker to be assigned right away. Other clients may begin to exhibit older, more problematic behaviors, as if to say, "You can't leave me; I still need you." Finally, clients may begin to devalue the work you have done together, since it is easier to say good-bye to something or someone who, after all, wasn't that important anyway.

For most interns, saying good-bye to clients is an emotional experience for them as well, sometimes in unexpected ways. You may find yourself feeling guilty that you haven't been able to do more for some or all of your clients, and again, this feeling is often unrelated to how much you have actually accomplished (Baird, 1996). In other cases, you may worry that you are the only one who can work successfully with a particular client or group of clients. It may be that you are the first person to really get through to a client. That is a wonderful feeling, but it places extra weight on the termination process. Although the client may regress some when you leave, chances are that at least some of the progress you have made will remain. It can, however, be hard to remember or believe that in the moment. If a client starts to regress, you may begin to question the value of your work or your effectiveness in general (Gould, 1978).

All human service workers are vulnerable to these reactions, but as an intern, especially if this is your first intensive experience in the helping field, you may be especially vulnerable. The closeness that can happen between worker and client is a heady, heartwarming experience, and it is especially affecting the first time (Baird, 1996). Both of us recall with special vividness our first few clients and the special difficulties that we faced in saying good-bye.

There are some pitfalls to watch out for as you take in and process your clients' reactions to your leaving. It is easy to personalize these reactions, as if they were statements about *you*. Of course, they have something to do with you, but remember that you may also be hearing expressions of their current struggles with themselves and echoes of old wounds. It is important to separate your clients' needs and issues from your own. If, for example, you struggle with a need to be liked, then you may be especially vulnerable to a client who turns on you in anger or devalues the experience. It is important to deal with these feelings and issues separately from the decisions you make about how to respond to your clients. You may need to be liked, but liking you in that moment may not be the best thing for your client. If a client has never been able to express disappointment with anyone, then the ability to do that with you is far more important than your need to be reassured. Baird (1996) also cautions that some interns unconsciously try to get clients to make the termination easier for them. They seek reassurance that they have done a good job or that whatever problems remain are not their fault. In these times, it is especially challenging to keep the client's needs in the foreground.

Another pitfall is promising to come back and visit. Clients may ask you to come back or to maintain contact in some other way. They may be avoiding the difficulties of separation, they may not understand the boundaries of a professional relationship, or they may just be doing what they would do with anyone they have come to care about. Even if they don't ask, you may find yourself wanting to reassure them that you will visit or keep in touch. Use extreme caution before making any such promises. Consult with your supervisor and your faculty instructor. Above all, if you do make commitments, make them with care; the worst thing you can do to some clients is promise to keep in touch and then fail to do so. As part of planning for termination, you need to work closely with your supervisor to set clear goals for yourself and your clients, and acknowledge clearly what your own needs and fears may be.

The Future

The last issue to think about in termination with your clients is the future, both for them and for your relationship with them. Just as this is a time to review goals and accomplishments, it is also a time to set or reestablish goals for the future. You also need to think about who will be dealing with your clients after you leave. In many agencies, intern supervisors plan for this transition, and you should be part of this planning. If your supervisor is not raising this issue, we suggest that you do. This transition, often referred to as *transferring clients*, requires time and attention if it is to be done well (Baird, 1996; Collins et al., 1992; Faiver, Eisengart, & Colonna, 1994).

The nature of the transfer will depend a great deal on your specific setting. You may actually have some clients that have been assigned to you. For example, there may be individuals who come to a clinic for counseling, a group you have been assigned to lead or colead, children and families whom you visit at their homes, or community groups that you meet with on a regular basis. In residential settings, it is not unusual for each resident to have a primary staff member and sometimes interns move into this capacity after they have been there for a few weeks. In other settings, there will not be clients who have been officially designated as "yours," but you may have developed especially close relationships with certain individuals, and it may be clear to you that you are the one to whom they turn most often. In any case, the issue of who is going to take over is one that should be discussed with your supervisor, with your coworkers, and of course with your clients. You should also be sure to reserve some time to work with the person who will be taking over, and you should begin this process well before your last day.

What sort of relationship will you have with your clients after you leave? What sort should you have? In many cases, the answer to both questions is "none." At most, you might keep in touch by mail. But for others, either because the client asks for more or because you want more, the issue becomes less clear. As we discussed in an earlier chapter, sexual relationships with clients while you are an intern are unacceptable and are a clear violation of the ethical standards of virtually every professional organization.

Social relationships with clients are also discouraged, although in some human service settings they are unavoidable and even beneficial. Now, however, you are leaving, and there may be clients with whom you want to pursue a friendship or a romance. Some of you, particularly those working with adolescents, may find this absurd and out of the question. But others are working with clients your own age and, trust us, the issue does come up.

This issue is complicated once again by the wide variety of placements and clients that interns deal with. Posttermination relationships have been the subject of a good deal of discussion in the counseling field. For example, Salisbury and Kiner (1996) conducted a survey of attitudes about friendship and sexual relationships with former clients. Of the counselors surveyed, 70% said they thought a friendship was acceptable, while 30% thought a sexual relationship could be acceptable. However, they also advised a lengthy waiting period, averaging 25 months for friendship and 62 months for romance. So if you were thinking that maybe you could just pick right up with a different kind of relationship with one of your clients, you may want to think again. Herlihey and Corey (1992) have provided a thoughtful review of these and many other issues about relationships with clients.

The major concern about posttermination relationships is the unfair power differential between you and your former client (Salisbury & Kiner, 1996). Through working with clients, human service workers often have knowledge of their most sensitive issues and areas of vulnerability. Clients, on the other hand, usually have no such knowledge of their workers, and this puts them in a vulnerable position in a friendship or romance.

However, not all of you are counselors, nor are you necessarily working with clients who have come to you or the agency because of psychological or emotional issues. If you are working in a recreation center, doing community organizing, or providing health education for college students, the power differential referred to may be quite different. So unfortunately, there are no hard and fast rules to follow, just some important issues to think about. If you are considering pursuing any sort of relationship with one of your clients, you ought to talk that over very carefully with your supervisor and your campus instructor. Whatever you decide, remember that even though these people may no longer be your clients, the primary concern in your decision making still must be their welfare, not yours.

SAYING GOOD-BYE TO YOUR SUPERVISOR

For most interns, the relationship with a supervisor is among the most significant at the internship. For many, the relationship has been close and positive, both intellectually and emotionally. For others, the relationship has not been as close or satisfying, but still one in which a lot was learned. Now that relationship is ending, and just as in your relationship with clients, the ending can engender feelings of a variety of natures and strengths for both of you. There is no way you are supposed to feel; interns report feelings ranging from mild to profound (Baird, 1996). If your relationship with your supervisor has been mostly an intellectual, dispassionate one, it may make saying good-bye easier (Haas, 1990). If you have not had a good relationship with your supervisor, the ending may be a relief. Regardless of how you feel, there are tasks you need to attend to in order to make the ending and the internship as productive as possible.

Coming to Closure: The Final Evaluation

You have probably had evaluations at your internship before this point; at least we hope so. Perhaps you are less nervous about them than you once were; perhaps not. In any case, you have one more to go through, and this can be the most important one of all. Many internship programs and placement sites use a written final evaluation of some sort. Most of them have their own format, so we are not going to provide one here. However, if you would like to see some examples, we suggest forms developed by Suanna Wilson (1981) or by Timothy Stanton and Kamil Ali (1994). As was the case with the initial or midsemester evaluation, we recommend that you become familiar with the form used at your placement or reach an agreement about what it will be before the evaluation is actually completed. That way you will have a better idea what to expect. You might even want to get an extra form and complete one about yourself. This can be a valuable process of reflection and insight.

Of equal if not greater importance is a final conference between you and your supervisor. This conference can either precede or follow the written evaluation. Some of the time can be used to prepare for that evaluation or to go over it. In any case, be sure to schedule an adequate block of time, at least an hour (Baird, 1996; Faiver et al., 1994; Shulman, 1983). There may be lots of tempting reasons for both you and your supervisor to avoid this session. You may be nervous about the feedback or about having to say good-bye. Your supervisor may not be good at endings either, and both of you may be pretty busy trying to wrap up projects or clients. This is one of those times when a formal, scheduled time—a ritual—will ensure that the task actually gets done.

The conference should cover both the cognitive and affective aspects of the ending; that is, it should deal with both thoughts and feelings. In the cognitive domain, Stanton and Ali (1994) have suggested dividing the conversation into two sections: work performance and learning. As you will see, these two areas have some overlap, but they are worth considering separately. In discussing work performance, you and your supervisor should cover the areas where you seem to have been especially effective, the clients with whom you have worked most successfully, and the service areas where you have demonstrated the most skill (Faiver et al., 1994). Both of your perceptions of these issues are important. In discussing what you have learned, you should use your written goals and objectives as a reference. You can prepare for this portion of the conference by reviewing the goals and rereading your journal. You may want to rate each of your goals and objectives.

The affective domain can be discussed separately or interwoven with the topics just mentioned. This is the time when you and your supervisor discuss your feelings about one another and about ending of the internship and your relationship. Some supervisors are more willing to share on this level than others; you may have reservations of your own. At the very least, be sure you are clear about how you feel. Then you can make a choice about how much to share with your supervisor.

It is important in this session to balance the positive and the critical (Baird, 1996; Shulman, 1983; Wilson, 1981). It is normal to want to focus on the positive and bask in the glow of your accomplishments and your supervisor's praise, but constructive criticism is just as important. There are bound to be things you need to work on and perhaps areas where you didn't perform as well as you or your supervisor would like. Just as your clients need to know where they still have work to do, so do you. Ignoring these areas is a little like saying, "I'm perfect. I learned everything I need to know to be a professional in this field in just a few short weeks." That sounds more like a quote from a late night TV advertisement than the stance of a reflective, thoughtful professional.

This conference will undoubtedly be more productive if you spend some time preparing for it. Get a clear idea of what will be discussed or what you would like to discuss. Reread your journal and make some notes for yourself. We also suggest you spend some time thinking about your emotional reactions to praise and criticism, as you did earlier. Some interns have a difficult time listening to criticism; others chafe at praise. Still others find it hard to tell other people how they really feel about them. Anything you can do to help yourself participate honestly and openly in this process will pay dividends.

The end of an internship may also be a time to offer some feedback to your supervisor (Baird, 1996; Collins et al., 1992; Faiver et al., 1994). Some supervisors will request this from you, and they may or may not give you advance notice. There are even forms available to offer written feedback (Collins et al., 1992). This is an opportunity to tell your supervisor what went well for you in the relationship and where there may be areas for improvement. It is important, though, to think carefully before offering criticism, regardless of how constructively put, unless it is requested (Baird, 1996). After all, your supervisor has and may continue to have power over you in the form of grades, letters of recommendation, and word of mouth in the community. Your campus instructor may be able to help you decide what and whether to share with your site supervisor.

One last issue to consider is a letter of recommendation. If you have had a good experience, you will probably want a letter from your supervisor, either for your next internship or for future employment. It's a good idea to get it now, while you and the internship are still fresh in the supervisor's mind, rather than call or write months later. Baird (1996) offers some important guidelines to follow when asking for a recommendation:

- Before requesting the letter, ask if your supervisor feels comfortable writing a supportive letter for you. This may seem silly, especially if you have a good relationship. Many supervisors, if they do not feel they can write a positive letter, will simply suggest that you get someone else. However, you don't want to take the chance that the letter will be less positive than you expected, especially if it is sent directly to a school or employer.
- Make clear what your future goals are. A letter for graduate school may look somewhat different than a letter for employment. There may also be some jobs for which your supervisor feels you are better qualified than others.

- Give plenty of notice and let your supervisor know if there is a deadline.
- Provide your supervisor with whatever forms and envelopes are needed. Pre-addressed and stamped envelopes show courtesy and consideration.

SAYING GOOD-BYE TO THE PLACEMENT

It is your last day, a day you have been looking forward to with an incredible mixture of feelings for weeks. You have had your final meeting with your supervisor and you have prepared your clients for your departure. You imagine what it will be like when everyone says good-bye to you and what you will say as you leave. As you move through the day, though, no one says a thing to you about leaving. Your supervisor doesn't seem to be around, and for your co-workers, it seems to be business as usual. Conversations focus on the work, and during quiet times, it's the usual office small-talk.

You go about your business, feeling a little confused but certain that someone will say something soon. Maybe they're being sneaky, and there is a little party planned for you later. No such luck. As you are leaving for the day, you say good-bye to people, pretty much as you always do. One of your co-workers looks up from what she's doing and says, "See you next week." "Well, no," you respond. "Today is my last day." "Oh my gosh," says your colleague. "I totally forgot! Well, hey, it's been great. Good luck to you." You start to reply, but the phone rings, a client calls, or a crisis erupts, and your colleague swirls away, back into the normal pace of life at the agency.

As you are driving home, you feel differently than you thought you would. Yes, there is joy and some relief, but also some hurt and a vague sense of emptiness. You try not to focus on that feeling, but it gnaws at you. What's wrong with those people anyway? Maybe you didn't mean as much to them as you thought. Maybe they're just rude and not quite the people you thought they were. You feel as if you deserve better from them after all you have done.

In a way, you're right. The events of the day have left you without a sense of appreciation and have interfered with your efforts to come to closure. Working with people for weeks and months and then forgetting their last day and not saying anything to them is perhaps insensitive. On the other hand, think about the pace of life at your placement. Most human service agencies are understaffed and extremely busy. In residential settings, large and small crises erupt all the time. And it's not the last day for anyone else. It may not even be the end of their day or week. So it is understandable that they might forget about your departure.

The point here is that you may need to be proactive in assuring you get that sense of closure you want. This is another time when a ritual may be in order. Some agencies have fairly elaborate rituals when someone leaves. There is a group meeting with clients, time set aside at a staff meeting, a party, or a farewell lunch. Other agencies don't do any of these things. Talk with your supervisor about the best way to mark the end of your internship. Ask what the norms are and be clear about what you need. It may be that there is just no time for a group activity, but at least you will know that, and you can schedule individual 15-minute sessions with some of your co-workers. In our experience, most supervisors are glad to help you make something happen, but they might not think of it on their own or may want to know what you prefer.

Even if there is going to be a formal good-bye celebration, there may be individual co-workers to whom you have grown especially close. They may take the initiative to have a final conversation with you, or they may not for any number of reasons. Here again, you need to take the initiative and schedule some time with them. This is yet another opportunity to practice your feedback skills; it is a time to let them know specifically what you have learned from them, what you appreciate about working with them, and how you feel about them. They may, in turn, do the same for you. Remember, though, the main idea is for you to say what you want and need to say, and if you do, the conversation is a success. You don't want to go away with that nagging feeling that you wish you had said such and such to so and so. Whatever you get back from them is an extra bonus.

EPILOGUE

Now your internship is over. Probably, there were times when you thought this day would never come or times when you were amazed and unnerved at how fast it was approaching. New challenges await you, of course. Some of you may be headed for another internship, some for a job in the field, and still others for the next level of schooling and internships.

It is important to remember that the stages of an internship are not phases you go through just once. In your new job or field placement, you are going to go through them again. The concerns about expectations and acceptance and the challenges of keeping yourself moving and growing, of confronting problems, and perhaps of ending well will all visit you again and again. When we tell our students this piece of news, some of them roll their eyes and hold their heads. But you have learned a great deal, and that learning will go with you.

You have learned valuable skills. If you continue to use them, they will grow sharper and more integrated over time. Even if your life and career change directions now, you can still take a lot of what you have learned with you. You have learned what to expect from an internship. The challenges will have a different shape and pace because they will be happening in a different place. You have grown and changed so much that it may feel like these challenges are happening to a different person. And of course, you will handle the issues in a new way. However, the concerns themselves should seem familiar to you if you can get enough distance and perspective from your everyday activities. Perhaps you will continue to keep a journal and pause for regular reflection on what you have written.

You have also learned a great deal about yourself. We have tried to emphasize that sort of learning in this book, and interns often tell us that this is the most powerful learning of all. You know more about how you respond to challenges and why. And you know that self-understanding is a process, not an accomplishment. You have more tools to pursue your self-understanding, and the practice you have had, if you continue it, will make those tools more and more second nature to you.

It may seem odd for us to say this, since we have never met you, but we wish you well. We are always glad to hear from former students and would be glad to hear from you about your continuing journey and how this book may or may not have been useful. We leave you with this quote from a student journal:

> I was given the opportunity to prove to myself that I could do it. This alone has allowed me to feel competent. I tested out my skills and got a professional feel about them. Now I have the key in my hand. I feel ready to move on. I am still not quite sure which doors this key will open, but I am sure that whatever I face I will deal with as best I know how.

For Further Reflection

1. What is on your agenda right now? What projects need to be finished at your internship? How about other classes? What else do you have to take care of in the next few weeks?
2. Think back to times in your life when you faced endings. What was it like for you then? What feelings did you have? How did you handle the ending?
3. Can you make any general statements about how you tend to handle good-byes? Are you satisfied with your patterns and tendencies in this area?
4. What will be most difficult in ending your internship? What can you do to ensure that you end the way you want to?
5. How will you approach termination with your clients? When will you tell clients you are leaving? How can you come to some closure with them? What future goals do you have for them? How do you feel about saying good-bye?
6. Think for a moment about your supervisor. What are some of the things you have learned? What is it that you most appreciate? What do you wish had been different? Is there anything you want to say to your supervisor before you leave?
7. Are there other individuals at your placement that you want to be sure to say good-bye to? How will you do this?
8. What plans do you have to acknowledge or celebrate the end of your internship?

For Further Exploration

Baird, B. N. (1996). *The internship, practicum and field placement handbook: A guide for the helping profession.* New Jersey: Prentice Hall.

This book contains a comprehensive, thoughtful discussion of a variety of issues associated with the ending of an internship.

Brill, N. (1998). *Working with people: The helping process.* (5th ed.). New York: Longman.

Especially helpful in thinking about termination with clients.

Gould, R. P. (1978). Students' experience with the termination phase of individual treatment. *Smith College Studies in Social Work, 48*(3), 235–269.

An excellent discussion of the unique nature and demands of "forced terminations."

References

Baird, B. N. (1996). *The internship, practicum, and field placement handbook: A guide for the helping professions.* Upper Saddle River, NJ: Prentice Hall.

Brill, N. (1998). *Working with people: The helping process* (5th ed.). New York: Longman.

Collins, D., Thomlison, B., & Grinnell, R. W. (1992). *The social work practicum: A student guide.* Itasca, IL: F. E. Peacock.

Fair, S. M., & Bressler, J. M. (1992). Therapist-initiated termination of psychotherapy. *The Clinical Supervisor, 10*(1), 171–189.

Faiver, C., Eisengart, S., & Colonna, R. (1994). *The counselor intern's handbook.* Pacific Grove, CA: Brooks/Cole.

Gould, R. P. (1978). Students' experience with the termination phase of individual treatment. *Smith College Studies in Social Work, 48*(3), 235–269.

Haas, S. (1990). *Hearing voices: Reflections of a psychology intern.* New York: Penguin.

Herlihey, B., & Corey, G. (1992). *Dual relationships in counseling.* Alexandria, VA: American Counseling Association.

Kegan, R. (1982). *The evolving self: Problem and process in human development.* Cambridge, MA: Harvard University Press.

Penn, L. S. (1990). When the therapist must leave: Forced termination of psychodynamic psychotherapy. *Professional Psychology: Research and Practice, 21*, 379–384.

Salisbury, W. A., & Kiner, R. T. (1996). Post termination friendship between counselors and clients. *Journal of Counseling and Development, 74*(5), 495–500.

Schafer, R. (1973). The termination of brief psychoanalytic psychotherapy. *International Journal of Psychoanalytic Psychotherapy, 11*, 135–148.

Shulman, L. (1983). *Teaching the helping skills: A field instructor's guide.* Itasca, IL: F. E. Peacock.

Stanton, T., & Ali, K. (1994). *The experienced hand: A student manual for making the most of an internship.* (2nd ed.). New York: Caroll Press.

Stanziani, P. (1993). *Practicum handbook: A guide to finding, obtaining, and getting the most out of an internship in the mental health field.* Cambridge, MA: Inky Publications.

Suelzle, M., & Borzak, L. (1981). Stages of fieldwork. In L. Borzak (Ed.), *Field study: A sourcebook for experiential learning* (pp. 136–150). Beverly Hills, CA: Sage Publications.

Watzlawick, P., Weaklund, J. H., & Fisch, F. (1974). *Change: Principles of problem formation and problem resolution.* New York: Norton.

Wilson, S. J. (1981). *Field instruction: Techniques for supervisors.* New York: Free Press.

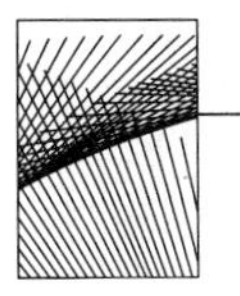

Index

TO THE OWNER OF THIS BOOK:

We hope that you have found *The Successful Internship: Transformation and Empowerment* useful. So that this book can be improved in a future edition, would you take the time to complete this sheet and return it? Thank you.

School and address: ______________________________

Department: ______________________________

Instructor's name: ______________________________

1. What I like most about this book is: ______________________________

____

____

2. What I like least about this book is: ______________________________

____

____

3. My general reaction to this book is: ______________________________

____

4. The name of the course in which I used this book is: ______________________________

____

5. Were all of the chapters of the book assigned for you to read? ______________

If not, which ones weren't? ______________________________

6. In the space below, or on a separate sheet of paper, please write specific suggestions for improving this book and anything else you'd care to share about your experience in using the book.

____

____

____

____

____

Optional:

Your name: ______________________ Date: ______________________

May Brooks/Cole quote you, either in promotion for *The Successful Internship: Transformation and Empowerment* or in future publishing ventures?

Yes: _________ No: _________

Sincerely,

H. Frederick Sweitzer
Mary A. King

FOLD HERE

FOLD HERE